M. Planck

TREATISE ON
THERMODYNAMICS

BY

DR. MAX PLANCK

PROFESSOR OF THEORETICAL PHYSICS IN THE UNIVERSITY OF BERLIN

TRANSLATED WITH THE AUTHOR'S SANCTION

BY

ALEXANDER OGG, M.A., B.Sc., Ph.D., F.Inst.P.

PROFESSOR OF PHYSICS, UNIVERSITY OF CAPETOWN, SOUTH AFRICA

THIRD EDITION
TRANSLATED FROM THE SEVENTH GERMAN EDITION

DOVER PUBLICATIONS, INC.
NEW YORK

This Dover edition, first published in 1969, is an
unabridged and unaltered republication of the
third revised edition originally published in 1917
by Mother Earth Publishing Association, New
York. A new Introduction by Richard Drinnon
has been added.

International Standard Book Number: 0-486-22484-8
Library of Congress Catalog Card Number: 72-98540

Manufactured in the United States by Courier Corporation
66371X07
www.doverpublications.com

PREFACE TO THE SIXTH AND THE SEVENTH EDITIONS.

AMONG the additions to the new edition, I may mention the theory of the lowering of the freezing point of strong electrolytes, which has been developed by J. C. Gosch of Calcutta. This theory (§ 273) appears at last to clear up the puzzling deviations from Ostwald's dilution law. Debye's equation of state for solid bodies has also been added. This contains the explanation of the variation of the specific heat with temperature as well as Grüneisen's law for the thermal coefficient of expansion.

THE AUTHOR.

BERLIN, GRUNEWALD,
February, 1921, and *July*, 1922.

TRANSLATOR'S NOTE.

THE Second English edition of Prof. Planck's book was a reprint of the first, but in this edition pains have been taken to adhere as closely as possible to the seventh German edition, with the exception of a few symbols, which have been retained as in the first English edition.

It has been a great pleasure to work up this edition containing all the additions and corrections, which Prof. Planck has found necessary. The value of the book has been greatly enhanced.

A. OGG.

UNIVERSITY OF CAPE TOWN,
January, 1926.

FROM THE PREFACE TO THE FIRST EDITION.

THE oft-repeated requests either to publish my collected papers on Thermodynamics, or to work them up into a comprehensive treatise, first suggested the writing of this book. Although the first plan would have been the simpler, especially as I found no occasion to make any important changes in the line of thought of my original papers, yet I decided to rewrite the whole subject-matter, with the intention of giving at greater length, and with more detail, certain general considerations and demonstrations too concisely expressed in these papers. My chief reason, however, was that an opportunity was thus offered of presenting the entire field of Thermodynamics from a uniform point of view. This, to be sure, deprives the work of the character of an original contribution to science, and stamps it rather as an introductory text-book on Thermodynamics for students who have taken elementary courses in Physics and Chemistry, and are familiar with the elements of the Differential and Integral Calculus.

The numerical values in the examples, which have been worked as applications of the theory, have, almost all of them, been taken from the original papers; only a few, that have been determined by frequent measurement, have been taken from the tables in Kohlrausch's "Leitfaden der praktischen Physik." It should be emphasized, however, that the numbers used, notwithstanding the care taken, have not undergone the same amount of critical sifting as the more general propositions and deductions.

———————

Three distinct methods of investigation may be clearly recognized in the previous development of Thermodynamics. The first penetrates deepest into the nature of the processes considered, and, were it possible to carry it out exactly, would be characterized as the most perfect. Heat, according to it, is due to the definite motions of the chemical molecules and atoms considered as distinct masses, which in the case of gases possess comparatively simple properties, but in the case of solids and liquids can be only very roughly sketched. This kinetic theory, founded by Joule, Waterston, Krönig and Clausius, has been greatly extended mainly by Maxwell and Boltzmann. Obstacles, at present unsurmountable, however, seem to stand in the way of its further progress. These are due not only to the highly complicated mathematical treatment, but principally to essential difficulties, not to be discussed here, in the mechanical interpretation of the fundamental principles of Thermodynamics.

Such difficulties are avoided by the second method, developed by Helmholtz. It confines itself to the most important hypothesis of the mechanical theory of heat, that heat is due to motion, but refuses on principle to specialize as to the character of this motion. This is a safer point of view than the first, and philosophically quite as satisfactory as the mechanical interpretation of nature in general, but it does not as yet offer a foundation of sufficient breadth upon which to build a detailed theory. Starting from this point of view, all that can be obtained is the verification of some general laws which have already been deduced in other ways direct from experience.

A third treatment of Thermodynamics has hitherto proved the most fruitful. This method is distinct from the other two, in that it does not advance the mechanical theory of heat, but, keeping aloof from definite assumptions as to its nature, starts direct from a few very general empirical facts, mainly the two fundamental principles of Thermodynamics.

From these, by pure logical reasoning, a large number of new physical and chemical laws are deduced, which are capable of extensive application, and have hitherto stood the test without exception.

This last, more inductive, treatment, which is used exclusively in this book, corresponds best to the present state of the science. It cannot be considered as final, however, but may have in time to yield to a mechanical, or perhaps an electro-magnetic theory. Although it may be of advantage for a time to consider the activities of nature—Heat, Motion, Electricity, etc.—as different in quality, and to suppress the question as to their common nature, still our aspiration after a uniform theory of nature, on a mechanical basis or otherwise, which has derived such powerful encouragement from the discovery of the principle of the conservation of energy, can never be permanently repressed. Even at the present day, a recession from the assumption that all physical phenomena are of a common nature would be tantamount to renouncing the comprehension of a number of recognized laws of interaction between different spheres of natural phenomena. Of course, even then, the results we have deduced from the two laws of Thermodynamics would not be invalidated, but these two laws would not be introduced as independent, but would be deduced from other more general propositions. At present, however, no probable limit can be set to the time which it will take to reach this goal.

BERLIN,
April, 1897.

FROM THE PREFACE TO THE SECOND EDITION.

THE numerous and valuable researches, which have enriched the subject of Thermodynamics since the publication of the first edition of this book, have added greatly to the abundance of known facts without in any way altering the fundamental basis of the theory. Since this book is chiefly concerned with the exposition of these fundamental principles, and the applications are given more as illustrative examples, I have not aimed at a new treatment of the subject, but have limited myself to corrections of some numerical data, and to a careful revision of the general ideas. I have thereby found it advisable to make a number of changes and additions. Many of these have been suggested by scientific acquaintances and colleagues.

With regard to the concluding paragraph of the preface to the first edition, I may be permitted to remark that the theory of heat has in the interval made remarkable progress along the path there indicated. Just as the first law of Thermodynamics forms only one side of the universal principle of the conservation of energy, so also the second law, or the principle of the increase of the Entropy, possesses no independent meaning. The new results of the investigations in the subject of heat radiation have made this still clearer. In this connection, I may mention the names of W. Wien, F. Paschen, O. Lummer and E. Pringsheim, H. Rubens, and F. Kurlbaum. The full content of the second law can only be understood if we look for its foundation in the known laws of the theory of probability as they were laid down by Clausius and Maxwell, and further extended by Boltzmann. According to this, the entropy of a natural state is in general equal to the logarithm of the *probability* of the corresponding state multiplied by a universal constant of the dimensions of energy divided by temperature. A closer discussion of this relation, which penetrates deeper than hitherto into a knowledge of molecular processes, as also of the laws of radiation, would overstep the limits which have been expressly laid down for this work. This discussion will therefore not be undertaken here, especially as I propose to deal with this subject in a separate book.

BERLIN,
January, 1905.

PREFACE TO THE THIRD EDITION.

THE plan of the presentation and the arrangement of the material is maintained in the new edition. Nevertheless, there is to be found in this edition, apart from a further revision of all the numerical data, a number of explanations and additions, which, one way or another, have been suggested. These are to be found scattered throughout the whole book. Of such I may mention, for example, the law of corresponding states, the definition of molecular weight, the proof of the second law, the characteristic thermodynamic function, the theory of the Joule–Thomson effect, and the evaporation of liquid mixtures. Further suggestions will always be very thankfully received.

A real extension of fundamental importance is the heat theorem, which was introduced by W. Nernst in 1906. Should this theorem, as at present appears likely, be found to hold good in all directions, then Thermodynamics will be enriched by a principle whose range, not only from the practical, but also from the theoretical point of view, cannot as yet be foreseen.

In order to present the true import of this new theorem in a form suitable for experimental test, it is, in my opinion, necessary to leave out of account its bearing on the atomic theory, which to-day is by no means clear. The methods, which have otherwise been adopted in this book, also depend on this point of view.

On the other hand, I have made the theorem, I believe, as general as possible, in order that its applications may be simple and comprehensive. Accordingly, Nernst's theorem has been extended both in form and in content. I mention this here as there is the possibility of the extended theorem not being confirmed, while Nernst's original theorem may still be true.

BERLIN,
November, 1910.

PREFACE TO THE FIFTH EDITION.

For the fifth edition, I have once more worked through
the whole material of the book, in particular the section on
Nernst's heat theorem. The theorem in its extended form
has in the interval received abundant confirmation and may
now be regarded as well established. Its atomic significance,
which finds expression in the restricted relations of the
quantum hypothesis, cannot, of course, be estimated in the
present work.

BERLIN,
March, 1917.

CONTENTS.

PART I.

FUNDAMENTAL FACTS AND DEFINITIONS.

PART II.

THE FIRST FUNDAMENTAL PRINCIPLE OF THERMODYNAMICS.

PART III.

THE SECOND FUNDAMENTAL PRINCIPLE OF THERMODYNAMICS.

PART IV.

APPLICATIONS TO SPECIAL STATES OF EQUILIBRIUM.

TREATISE

ON

THERMODYNAMICS.

PART I.

FUNDAMENTAL FACTS AND DEFINITIONS.

CHAPTER I.

TEMPERATURE.

§ 1. THE conception of " heat " arises from that particular
sensation of warmth or coldness which is immediately experi-
enced on touching a body. This direct sensation, however,
furnishes no quantitative scientific measure of a body's state
with regard to heat; it yields only qualitative results, which
vary according to external circumstances. For quantitative
purposes we utilize the change of volume which takes place
in all bodies when heated under constant pressure, for this
admits of exact measurement. Heating produces in most
substances an increase of volume, and thus we can tell whether
a body gets hotter or colder, not merely by the sense of touch,
but also by a purely mechanical observation affording a much
greater degree of accuracy. We can also tell accurately
when a body assumes a former state of heat.

§ 2. If two bodies, one of which feels warmer than the
other, be brought together (for example, a piece of heated
metal and cold water), it is invariably found that the hotter
body is cooled, and the colder one is heated up to a certain

point, and then all change ceases.   The two bodies are then said to be in *thermal equilibrium*.   Experience shows that such a state of equilibrium finally sets in, not only when two, but also when any number of differently heated bodies are brought into mutual contact.   From this follows the important proposition : *If a body*, A, *be in thermal equilibrium with two other bodies*, B *and* C, *then* B *and* C *are in thermal equilibrium with one another*.   For, if we bring A, B, and C together so that each touches the other two, then, according to our supposition, there will be equilibrium at the points of contact AB and AC, and, therefore, also at the contact BC.   If it were not so, no general thermal equilibrium would be possible, which is contrary to experience.

§ **3.**   These facts enable us to compare the degree of heat of two bodies, B and C, without bringing them into contact with one another;   namely, by bringing each body into contact with an arbitrarily selected standard body, A (for example, a mass of mercury enclosed in a vessel terminating in a fine capillary tube).   By observing the volume of A in each case, it is possible to tell whether B and C are in thermal equilibrium or not.   If they are not in thermal equilibrium, we can tell which of the two is the hotter.   The degree of heat of A, or of any body in thermal equilibrium with A, can thus be very simply defined by the volume of A, or, as is usual, by the difference between the volume of A and an arbitrarily selected *normal volume*, namely, the volume of A when in thermal equilibrium with melting ice under atmospheric pressure.   This volumetric difference, which, by an appropriate choice of unit, is made to read 100 when A is in contact with steam under atmospheric pressure, is called the *temperature* in degrees Centigrade with regard to A as thermometric substance.   Two bodies of equal temperature are, therefore, in thermal equilibrium, and *vice versâ*.

§ **4.**   The temperature readings of no two thermometric substances agree, in general, except at 0° and 100°   The definition of temperature is therefore somewhat arbitrary. This we may remedy to a certain extent by taking gases, in particular those hard to condense, such as hydrogen, oxygen, nitrogen, and carbon monoxide, and all so-called permanent gases as thermometric substances. They agree almost completely within a considerable range of temperature, and their readings are sufficiently in accordance for most purposes. Besides, the coefficient of expansion of these different gases is the same, inasmuch as equal volumes of them expand under constant pressure by the same amount—about $\frac{1}{273}$ of their volume—when heated from 0° C. to 1° C. Since, also, the influence of the external pressure on the volume of these gases can be represented by a very simple law, we are led to the conclusion that these regularities are based on a remarkable simplicity in their constitution, and that, therefore, it is reasonable to define the common temperature given by them simply as temperature. We must consequently reduce the readings of other thermometers to those of the gas thermometer.

§ 5. The definition of temperature remains arbitrary in cases where the requirements of accuracy cannot be satisfied by the agreement between the readings of the different gas thermometers, for there is no sufficient reason for the preference of any one of these gases. A definition of temperature completely independent of the properties of any individual substance, and applicable to all stages of heat and cold, becomes first possible on the basis of the *second law of thermodynamics* (§ 160, etc.). In the mean time, only such temperatures will be considered as are defined with sufficient accuracy by the gas thermometer.

§ 6. In the following we shall deal chiefly with homogeneous, isotropic bodies of any form, possessing throughout their substance the same temperature and density, and subject to a uniform pressure acting everywhere perpendicular to the surface. They, therefore, also exert the same pressure outwards. Surface phenomena are thereby disregarded. The condition of such a body is determined by its chemical nature; its mass, M; its volume, V; and its temperature, t. On these must depend, in a definite manner, all other properties of the particular state of the body, especially the pressure, which is uniform throughout, internally and externally. The pressure, p, is measured by the force acting on the unit of area—in the c.g.s. system, in dynes per square centimeter, a *dyne* being the force which imparts to a mass of one gramme in one second a velocity of one centimeter per second.

§ 7. As the pressure is generally given in atmospheres, the value of an atmosphere in absolute C.G.S. units is here calculated. The pressure of an atmosphere is the force which a column of mercury at 0° C., 76 cm. high, and 1 sq. cm. in cross-section exerts on its base in consequence of its weight, when placed in geographical latitude 45°. This latter condition must be added, because the weight, *i.e.* the force of the earth's attraction, varies with the locality. The volume of the column of mercury is 76 c.c.; and since the density of mercury at 0° C. is 13·596 $\frac{\text{grm.}}{\text{cm.}^3}$, the mass is 76 × 13·596 grm. Multiplying the mass by the acceleration of gravity in latitude 45°, we find the pressure of one atmosphere in absolute units to be

$$76 \times 13 \cdot 596 \times 980 \cdot 6 = 1,013,250 \, \frac{\text{dyne}}{\text{cm.}^2} \text{ or } \frac{\text{gr.}}{\text{cm.-sec.}^2}.$$

If, as was formerly the custom in mechanics, we use as the unit of force the weight of a gramme in geographical latitude 45° instead of the dyne, the pressure of an atmosphere would be 76 × 13·596 = 1033·3 grm. per square centimeter.

§ 8. Since the pressure in a given substance is evidently controlled by its internal physical condition only, and not by its form or mass, it follows that p depends only on the temperature and the ratio of the mass M to the volume V (*i.e.* the density), or on the reciprocal of the density, the volume of unit mass :

$$\frac{V}{M} = v,$$

which is called the specific volume of the substance. For every substance, then, there exists a characteristic relation—

$$p = f(v,t),$$

which is called the *characteristic equation* of the substance. For gases, the function f is invariably positive; for liquids and solids, however, it may have also negative values under certain circumstances.

§ 9. **Perfect Gases.**—The characteristic equation assumes its simplest form for the substances which we used in § 4 for the definition of temperature, and in so far as they yield corresponding temperature data are called *ideal* or *perfect* gases. If the temperature be kept constant, then, according to the Boyle–Mariotte law, the product of the pressure and the specific volume remains constant for gases :

$$pv = \theta, \quad \ldots \quad \ldots \quad (1)$$

where θ, for a given gas, depends only on the temperature.

But if the pressure be kept constant, then, according to § 3, the temperature is proportional to the difference between the present volume v and the normal volume v_0; *i.e.* :

$$t = (v - v_0)P, \quad \ldots \quad \ldots \quad (2)$$

where P depends only on the pressure p. Equation (1) becomes

$$pv_0 = \theta_0, \quad \ldots \quad \ldots \quad (3)$$

where θ_0 is the value of the function θ, when $t = 0°$ C.

Finally, as has already been mentioned in § 4, the expansion of all permanent gases on heating from 0° C. to 1° C. is the same fraction α (about $\frac{1}{273}$) of their volume at 0° (Gay

Lussac's law). Putting $t = 1$, we have $v - v_0 = \alpha v_0$, and equation (2) becomes

$$1 = \alpha v_0 \mathrm{P} \quad . \quad . \quad . \quad . \quad . \quad (4)$$

By eliminating P, v_0, and v from (1), (2), (3), (4), we obtain the temperature function of the gas—

$$\theta = \theta_0(1 + \alpha t),$$

which is seen to be a linear function of t. The characteristic equation (1) becomes

$$p = \frac{\theta_0}{v}(1 + \alpha t).$$

§ 10. The form of this equation is considerably simplified by shifting the zero of temperature, arbitrarily fixed in § 3, by $\dfrac{1}{\alpha}$ degrees, and calling the melting point of ice, not 0° C., but $\dfrac{1°}{\alpha}$ C. (*i.e.* about 273° C.). For, putting $t + \dfrac{1}{\alpha} = \mathrm{T}$ (absolute temperature), and the constant $\alpha v_0 = \mathrm{C}$, the characteristic equation becomes

$$p = \frac{\mathrm{C}}{v}\mathrm{T} = \frac{\mathrm{CM}}{\mathrm{V}}\mathrm{T} \quad . \quad . \quad . \quad . \quad (5)$$

This introduction of *absolute* temperature is evidently tantamount to measuring temperature no longer, as in § 3, by a change of volume, but by the volume itself.

The question naturally arises, What is the physical meaning of the zero of absolute temperature? The zero of absolute temperature is that temperature at which a perfect gas of finite volume has no pressure, or under finite pressure has no volume. This statement, when applied to actual gases, has no meaning, since by requisite cooling they show considerable deviations from one another and from the ideal state. How far an actual gas by average temperature changes deviates from the ideal cannot of course be tested, until temperature has been defined without reference to any particular substance (§ 5).

§ 11. The constant C, which is characteristic for the perfect gas under consideration, can be calculated, if the specific volume v be known for any pair of values of T and p (*e.g.* 0° C. and 1 atmosphere). For different gases, taken at the same temperature and pressure, the constants C evidently vary directly as the specific volumes, or inversely as the densities $\frac{1}{v}$. It may be affirmed, then, that, taken at the same temperature and pressure, the densities of all perfect gases bear a constant ratio to one another. A gas is, therefore, often characterized by the constant ratio which its density bears to that of a normal gas at the same temperature and pressure (*specific density* relative to air or hydrogen). At 0° C. (T = 273°) and under 1 atmosphere pressure, the densities of the following gases are :

Hydrogen	$0.00008988 \frac{\text{gr.}}{\text{cm.}^3}$
Oxygen	0·0014291
Nitrogen	0·0012507
Atmospheric nitrogen	0·0012567
Air	0·0012928
Argon	0·0017809

whence the corresponding values of C in absolute units can be readily calculated.

All questions with regard to the behaviour of a substance when subjected to changes of temperature, volume, and pressure are completely answered by the characteristic equation of the substance.

§ 12. Behaviour under Constant Pressure (Isobaric or Isopiestic Changes).—*Coefficient of expansion* is the name given to the ratio of the increase of volume for a rise of temperature of 1° C. to the volume at 0° C., *i.e.* the quantity $\frac{V_{T+1} - V_T}{V_0}$. Since as a rule the volume changes comparatively slowly with temperature we may put it $= \left(\frac{\partial V}{\partial T}\right)_p \cdot \frac{1}{V_0}$.

For a perfect gas according to equation (5) $V_{T+1} - V_T = \frac{CM}{p}$

and $V_0 = \dfrac{CM}{p} \times 273$. The coefficient of expansion of the gas is then $\frac{1}{273} = \alpha$.

§ 13. Behaviour at Constant Volume (Isochoric or Isopycnic or Isosteric Changes).

—The *pressure coefficient* is the ratio of the increase of pressure for a rise of temperature of 1° C. to the pressure at 0° C., *i.e.* the quantity $\dfrac{p_{T+1} - p_T}{p_0}$, or $\left(\dfrac{\partial p}{\partial T}\right)_V \times \dfrac{1}{p_0}$. For an ideal gas, according to equation (5),

$p_{T+1} - p_T = \dfrac{CM}{V}$, and $p_0 = \dfrac{CM}{V} \times 273$. The pressure coefficient of the gas is $\frac{1}{273}$, equal to the coefficient of expansion α.

§ 14. Behaviour at Constant Temperature (Isothermal Changes).

—*Coefficient of elasticity* is the ratio of an infinitely small increase of pressure to the resulting contraction of unit volume of the substance, *i.e.* the quantity

$$-\frac{dp}{\dfrac{dV}{V}} = -\left(\frac{\partial p}{\partial v}\right)_T v.$$

For an ideal gas, according to equation (5),

$$\left(\frac{\partial p}{\partial v}\right)_T = -\frac{CT}{v^2}.$$

The coefficient of elasticity of the gas is, therefore,

$$\frac{CT}{v} = p,$$

that is, equal to the pressure.

The reciprocal of the coefficient of elasticity, *i.e.* the ratio of an infinitely small contraction of unit volume to the corresponding increase of pressure, $-\left(\dfrac{\partial v}{\partial p}\right)_T \cdot \dfrac{1}{v}$, is called the *coefficient of compressibility*.

§ 15. The three coefficients which characterize the behaviour of a substance subject to isobaric, isochoric, and isothermal changes are not independent of one another, but are in every case connected by a definite relation. The general characteristic equation, on being differentiated, gives

$$dp = \left(\frac{\partial p}{\partial T}\right)_v dT + \left(\frac{\partial p}{\partial v}\right)_T dv,$$

where the suffixes indicate the variables to be kept constant while performing the differentiation. By putting $dp = 0$ we impose the condition of an isobaric change, and obtain the relation between dv and dT in isobaric processes :—

$$\left(\frac{\partial v}{\partial T}\right)_p = -\frac{\left(\frac{\partial p}{\partial T}\right)_v}{\left(\frac{\partial p}{\partial v}\right)_T} . \quad \ldots \quad (6)$$

For every state of a substance, one of the three coefficients, viz. of expansion, of pressure, or of compressibility, may therefore be calculated from the other two.

Take, for example, mercury at 0° C. and under atmospheric pressure. Its coefficient of expansion is (§ 12)

$$\left(\frac{\partial v}{\partial T}\right)_p \cdot \frac{1}{v_0} = 0{\cdot}00018,$$

its coefficient of compressibility in atmospheres (§ 14) is

$$-\left(\frac{\partial v}{\partial p}\right)_T \cdot \frac{1}{v_0} = 0{\cdot}0000039,$$

therefore its pressure coefficient in atmospheres (§ 13) is

$$\left(\frac{\partial p}{\partial T}\right)_v = -\left(\frac{\partial p}{\partial v}\right)_T \cdot \left(\frac{\partial v}{\partial T}\right)_p = -\frac{\left(\frac{\partial v}{\partial T}\right)_p}{\left(\frac{\partial v}{\partial p}\right)_T} = \frac{0{\cdot}00018}{0{\cdot}0000039} = 46.$$

This means that an increase of pressure of 46 atmospheres

is required to keep the volume of mercury constant when heated from 0° C. to 1° C.

§ 16. **Mixtures of Perfect Gases.**—If any quantities of the *same* gas at the same temperatures and pressures be at first separated by partitions, and then allowed to come suddenly in contact with another by the removal of these partitions, it is evident that the volume of the entire system will remain the same and be equal to the sum-total of the partial volumes. Starting with quantities of *different* gases, experience still shows that, when pressure and temperature are maintained uniform and constant, the total volume continues equal to the sum of the volumes of the constituents, notwithstanding the slow process of intermingling—diffusion —which takes place in this case. Diffusion goes on until the mixture has become at every point of precisely the same composition, *i.e.* physically homogeneous.

§ 17. Two views regarding the constitution of mixtures thus formed present themselves. Either we might assume that the individual gases, while mixing, split into a large number of small portions, all retaining their original volumes and pressures, and that these small portions of the different gases, without penetrating each other, distribute themselves evenly throughout the entire space. In the end each gas would still retain its original volume (partial volume), and all the gases would have the same common pressure. Or, we might suppose—and this view will be shown below (§ 32) to be the correct one—that the individual gases change and interpenetrate in every infinitesimal portion of the volume, and that after diffusion each individual gas, in so far as one may speak of such, fills the total volume, and is consequently under a lower pressure than before diffusion. This so-called partial pressure of a constituent of a gas mixture can easily be calculated.

§ 18. Denoting the quantities referring to the individual gases by suffixes—T and p requiring no special designation, as they are supposed to be the same for all the gases—the characteristic equation (5) gives for each gas before diffusion

$$p = \frac{C_1 M_1 T}{V_1}; \; p = \frac{C_2 M_2 T}{V_2}; \; \ldots$$

The total volume,

$$V = V_1 + V_2 + \ldots,$$

remains constant during diffusion. After diffusion we ascribe to each gas the total volume, and hence the partial pressures become

$$p_1 = \frac{C_1 M_1 T}{V} = \frac{V_1}{V}p; \; p_2 = \frac{C_2 M_2 T}{V} = \frac{V_2}{V}p; \; \ldots \quad (7)$$

and by addition

$$p_1 + p_2 + \ldots = \frac{V_1 + V_2 + \ldots}{V}p = p \quad . \quad . \quad (8)$$

This is Dalton's law, that in a homogeneous mixture of gases the pressure is equal to the sum of the partial pressures of the gases. It is also evident that

$$p_1 : p_2 : \ldots = V_1 : V_2 : \ldots = C_1 M_1 : C_2 M_2 \quad . \quad . \quad (9)$$

i.e. the partial pressures are proportional to the volumes of the gases before diffusion, or to the partial volumes which the gases would have according to the first view of diffusion given above.

§ **19.** The characteristic equation of the mixture, according to (7) and (8), is

$$p = (C_1 M_1 + C_2 M_2 + \ldots)\frac{T}{V}$$

$$= \left(\frac{C_1 M_1 + C_2 M_2 + \ldots}{M}\right)\frac{M}{V}T \quad . \quad . \quad (10)$$

which corresponds to the characteristic equation of a perfect gas with the following characteristic constant :

$$C = \frac{C_1 M_1 + C_2 M_2 + \ldots}{M_1 + M_2 + \ldots} \quad . \quad . \quad . \quad (11)$$

Hence the question as to whether a perfect gas is a chemically simple one, or a mixture of chemically different gases, cannot in any case be settled by the investigation of the characteristic equation.

§ 20. The composition of a gas mixture is defined, either by the ratios of the masses, M_1, M_2, . . . or by the ratios of the partial pressures p_1, p_2, . . . or the partial volumes V_1, V_2, . . . of the individual gases. Accordingly we speak of per cent. by weight or by volume. Let us take for example atmospheric air, which is a mixture of oxygen (1) and " atmospheric " nitrogen (2).

The ratio of the densities of oxygen, " atmospheric " nitrogen and air is, according to § 11,

$$0\cdot 0014291 : 0\cdot 0012567 : 0\cdot 0012928 = \frac{1}{C_1} : \frac{1}{C_2} : \frac{1}{C}.$$

Taking into consideration the relation (11) :

$$C = \frac{C_1 M_1 + C_2 M_2}{M_1 + M_2},$$

we find the ratio

$$M_1 : M_2 = \frac{C_2 - C}{C - C_1} = 0\cdot 3009,$$

i.e. 23·1 per cent. by weight of oxygen and 76·9 per cent. of nitrogen. On the other hand, the ratio

$$C_1 M_1 : C_2 M_2 = p_1 : p_2 = V_1 : V_2 = \frac{\dfrac{1}{C} - \dfrac{1}{C_2}}{\dfrac{1}{C_1} - \dfrac{1}{C}} = 0\cdot 2649,$$

i.e. 20·9 per cent. by volume of oxygen and 79·1 per cent. of nitrogen.

§ 21. **Characteristic Equation of Other Substances.**—The characteristic equation of perfect gases, even in the case of the substances hitherto discussed, is only an approximation, though a close one, to the actual facts. A still further deviation from the behaviour of perfect gases is shown by the other gaseous bodies, especially by those easily condensed, which for this reason were formerly classed as *vapours*. For these a modification of the characteristic equation is necessary. It is worthy of notice, however, that the more rarefied the state in which we observe these gases, the less does their behaviour deviate from that of perfect gases, so that all gaseous substances, when sufficiently rarefied, may be said in general to act like perfect gases even at low temperatures. The general characteristic equation of gases and vapours, for very large values of v, will pass over, therefore, into the special form for perfect gases.

§ 22. We may obtain by various graphical methods an idea of the character and magnitude of the deviations from the ideal gaseous state. An isothermal curve may, *e.g.*, be drawn, taking v and p for some given temperature as the abscissa and ordinate, respectively, of a point in a plane. The entire system of isotherms gives us a complete representation of the characteristic equation. The more the behaviour of the vapour in question approaches that of a perfect gas, the closer do the isotherms approach those of equilateral hyperbolæ having the rectangular co-ordinate axes for asymptotes, for $pv =$ const. is the equation of an isotherm of a perfect gas. The deviation from the hyperbolic form yields at the same time a measure of the departure from the ideal state.

§ 23. The deviations become still more apparent when the isotherms are drawn taking the product pv (instead of p) as the ordinate and say p as the abscissa. Here a perfect gas has evidently for its isotherms straight lines parallel to the axis of abscissæ. In the case of actual gases, however, the isotherms slope gently towards a minimum value of pv, the position of which depends on the temperature and the nature of the gas. For lower pressures (*i.e.* to the left of the minimum), the volume decreases at a more rapid rate, with increasing pressure, than in the case of perfect gases; for higher pressures (to the right of the minimum), at a slower rate. At the minimum point the compressibility coincides with that of a perfect gas. In the case of hydrogen the minimum lies far to the left, and it has hitherto been possible to observe it only at very low temperatures.

§ 24. To van der Waals is due the first analytical formula for the general characteristic equation, applicable also to the liquid state. He also explained physically, on the basis of the kinetic theory of gases, the deviations from the behaviour of perfect gases. As we do not wish to introduce here the hypothesis of the kinetic theory, we consider van der Waals' equation merely as an approximate expression of the facts. His equation is

$$p = \frac{RT}{v - b} - \frac{a}{v^2}, \quad . \quad . \quad . \quad . \quad (12)$$

where R, a, and b are constants which depend on the nature of the substance. For large values of v, the equation, as required, passes into that of a perfect gas; for small values of v and corresponding values of T, it represents the characteristic equation of a liquid.

Expressing p in atmospheres and calling the specific volume v unity for T = 273 and $p = 1$, van der Waals' constants for carbon dioxide are

R = 0·00369; a = 0·00874; b = 0·0023.

As the volume of 1 gr. of carbon dioxide at 0° C. and atmospheric pressure is 506 c.c., the values of v calculated from the formula must be multiplied by 506 to obtain the specific volumes in absolute units.

§ 25. Van der Waals' equation not being sufficiently accurate, Clausius supplemented it by the introduction of additional constants. Clausius' equation is

$$p = \frac{RT}{v - a} - \frac{c}{T(v + b)^2} \quad . \quad . \quad . \quad (12a)$$

For large values of v, this too approaches the characteristic equation of an ideal gas. In the same units as above, Clausius' constants for carbon dioxide are :

$$R = 0·003688; \quad a = 0·000843; \quad b = 0·000977; \quad c = 2·0935.$$

Observations on the compressibility of gaseous and liquid carbon dioxide at different temperatures are fairly well satisfied by Clausius' formula.

Many other forms of the characteristic equation have been deduced by different scientists, partly on experimental and partly on theoretical grounds. A very useful formula for gases at not too high pressures was given by D. Berthelot.

§ 26. If, with p and v as ordinates, we draw the isotherms representing Clausius' equation for carbon dioxide, we obtain the graphs of Fig. 1.*

For high temperatures the isotherms approach equilateral hyperbolæ, as may be seen from equation (12a). In general, however, the isotherm is a curve of the third degree, three values of v corresponding to one of p. Hence, in general, a straight line parallel to the axis of abscissæ intersects an isotherm in three points, of which two, as actually happens for large values of T, may be imaginary. At high temperatures there is, consequently, only one real volume corresponding to a given pressure, while at lower temperatures, there are three real values of the volume for a given pressure. Of these three values (indicated on the figure by α, β, γ, for instance) only the smallest (α) and the largest (γ) represent practically realizable states, for at the middle point (β) the pressure along the isotherm would increase with increasing volume, and the compressibility would accordingly be negative. Such a state has, therefore, only a theoretical signification.

§ 27. The point α corresponds to liquid carbon dioxide, and γ to the gaseous condition at the temperature of the isotherm passing through the points and under the pressure measured by the ordinates of the line $\alpha\beta\gamma$. In general only

* For the calculation and construction of the curves, I am indebted to Dr. Richard Apt.

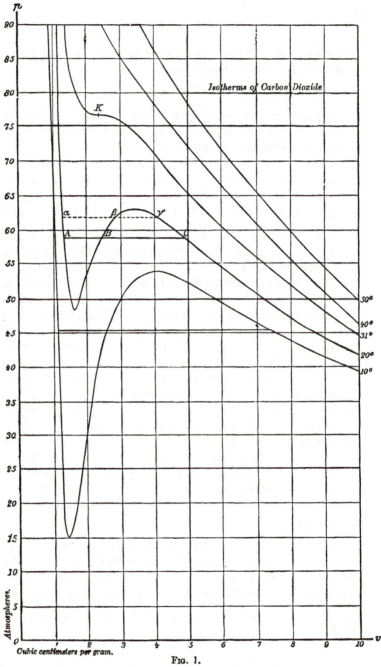

FIG. 1.

ISOTHERMS OF CARBON DIOXIDE FROM CLAUSIUS' EQUATION.

one of these states is stable (in the figure, the liquid state at α). For, if we compress gaseous carbon dioxide, enclosed in a cylinder with a movable piston, at constant temperature, *e.g.* at 20° C., the gas assumes at first states corresponding to consecutive points on the 20° isotherm to the extreme right. The point representative of the physical state of the gas then moves farther and farther to the left until it reaches a certain place C. After this, further compression does not move the point beyond C, but there now takes place a partial condensation of the substance—a splitting into a liquid and a gaseous portion. Both parts, of course, possess common pressure and temperature. The state of the gaseous portion continues to be characterized by the point C, that of the liquid portion by the point A of the same isotherm. C is called the saturation point of carbon dioxide gas for the particular temperature considered. Isothermal compression beyond C merely results in precipitating more of the vapour in liquid form. During this part of the isothermal compression no change takes place. but the condensation of more and more vapour; the internal conditions (pressure, temperature, specific volume) of both parts of the substance are always represented by the two points A and C. At last, when all the vapour has been condensed, the whole substance is in the liquid condition A, and again behaves as a homogeneous substance, so that further compression gives an increase of density and pressure along the isotherm. The substance will now pass through the point α of the figure. On this side, as may be seen from the figure, the isotherm is much steeper than on the other, *i.e.* the compressibility is much smaller.

At times, it is possible to follow the isotherm beyond the point C towards the point γ, and to prepare a so-called supersaturated vapour. Then only a more or less unstable condition of equilibrium is obtained, as may be seen from the fact that the smallest disturbance of the equilibrium is sufficient to cause an immediate condensation. The substance passes by a jump into the stable condition. Nevertheless, by

the study of supersaturated vapours, the theoretical part of
the curve also receives a direct meaning.

§ **28.** On any isotherm, which for certain values of p
admits of three real values of v, there are, therefore, two
definite points, A and C, corresponding to the state of saturation. The position of these points is not immediately deducible
from the graph of the isotherm. The propositions of thermodynamics, however, lead to a simple way of finding these
points, as will be seen in § 172. The higher the temperature,
the smaller becomes the region in which lines drawn parallel
to the axis of abscissæ intersect the isotherm in three real
points, and the closer will these three points approach one
another. The transition to the hyperbola-like isotherms,
which any parallel to the axis of abscissæ cuts in one point
only, is formed by that particular isotherm on which the
three points of intersection coalesce into one, giving a point
of inflection. The tangent to the curve at this point is
parallel to the axis of abscissæ. It is called the *critical
point* (K of Fig. 1) of the substance, and its position indicates
the critical temperature, the critical specific volume, and the
critical pressure of the substance. At this point there is no
difference between the saturated vapour and its liquid precipitate. Above the critical temperature and critical pressure,
condensation does not exist, as the diagram plainly shows.
Hence all attempts to condense hydrogen, oxygen, and
nitrogen necessarily failed as long as the temperature had not
been reduced below the critical temperature, which is very
low for these gases.

§ **29.** It further appears from the figure that there is no
definite boundary between the gaseous and liquid states,
since from the region of purely gaseous states, as at C, that
of purely liquid ones, as at A, may be reached on a circuitous
path that nowhere passes through a state of saturation—on
a curve, for instance, drawn around the critical point. Thus
a vapour may be heated at constant volume above the critical
temperature, then compressed at constant temperature below

the critical volume, and finally cooled under constant pressure below the critical temperature. Condensation nowhere occurs in this process, which leads, nevertheless, to a region of purely liquid states. The earlier fundamental distinction between liquids, vapours, and gases should therefore be dropped as no longer tenable. A more modern proposal to denote as gaseous all states above the critical temperature, and as vaporous or liquid all others according as they lie to the right or left of the theoretical regions (Fig. 1), has also this disadvantage, that thereby a boundary is drawn between liquid and gas on the one hand, and vapour and gas on the other hand, which has no physical meaning. The crossing of the critical temperature at a pressure other than the critical pressure differs in no way from the crossing of any other temperature.

§ **30.** The position of the critical point may be readily calculated from the general characteristic equation. According to § 28 we have

$$\left(\frac{\partial p}{\partial v}\right)_{\mathrm{T}} = 0, \text{ and } \left(\frac{\partial^2 p}{\partial v^2}\right)_{\mathrm{T}} = 0.$$

The first of these means that the tangent to the isotherm at K is parallel to the axis of abscissæ; and the second, that the isotherm has a point of inflection at K. If we choose van der Waals' characteristic equation (12), we find for the critical point

$$\mathrm{T}_c = \frac{8a}{27b\mathrm{R}}, \quad p_c = \frac{a}{27b^2}, \quad v_c = 3b.$$

From the numbers already given, the critical constants of carbon dioxide are

$$\mathrm{T}_c = 305 = 273° + 32°, \quad p_c = 61\cdot2 \text{ atmos.}, \quad v_c = 0\cdot0069.$$

In c.g.s. units $v_c = 0\cdot0069 \times 506 = 3\cdot49$ c.c./grm. Instead of the three constants a, b, and R, we may introduce the three constants T_c, p_c, and v_c into the equation. Since the units in which the temperature, the pressure, and the volume are measured are quite independent of one another, it follows that only the ratios of T and T_c, p and p_c, v and v_c occur in the characteristic equation. In other words, if we take the ratios,

$$\frac{T}{T_c} = \tau, \ \frac{p}{p_c} = \pi, \ \frac{v}{v_c} = \phi$$

(*reduced* temperature, *reduced* pressure, and *reduced* volume), as independent variables, the characteristic equation contains no constant depending on the special nature of the substance. In fact, we find in this way for van der Waals' equation of state

$$\pi = \frac{8\tau}{3\phi - 1} - \frac{3}{\phi^2}.$$

When $\tau = 1$ and $\phi = 1$, then $\pi = 1$.

This law that the characteristic equation, when expressed in terms of reduced values of the temperature, pressure, and volume, is the same for all gases, is called the *law of corresponding states.* It is not only true of van der Waals' equation, but of every equation of state in which there are not more than three constants depending on the nature of the substance. The law, however, is only approximately true, which from the point of view of the molecular theory is illuminating, since there is no single substance which in a chemical sense remains absolutely the same under all temperature and volume changes (see § 33).

§ **30A.** If we take Clausius' equation with four constants, we obtain for the critical point,

$$T_c{}^2 = \frac{8c}{27(a+b)R}, \ p_c{}^2 = \frac{cR}{216(a+b)^3}, \ v_c = 3a + 2b.*$$

* Obtained as follows :

$$p = \frac{RT}{v - a} - \frac{c}{T(v + b)^2} \quad \cdot \quad \cdot \quad \cdot \quad \cdot \quad (1)$$

$$\left(\frac{\partial p}{\partial v}\right)_T = -\frac{RT}{(v - a)^2} + \frac{2c}{T(v + b)^3} = 0 \quad \cdot \quad \cdot \quad (2)$$

$$\left(\frac{\partial^2 p}{\partial v^2}\right)_T = \frac{2RT}{(v - a)^3} - \frac{6c}{T(v + b)^4} = 0 \quad \cdot \quad \cdot \quad (3)$$

The critical constants of carbon dioxide are then,

$$T_c = 304 = 273° + 31°, \quad p_c = 77 \text{ atmos.}, \quad v_c = 2\cdot27 \, \frac{\text{c.c.}}{\text{gr.}}.$$

These values lie considerably nearer the experimental values $T_c = 273° + 31°$, $p_c = 73$ atmos., $v_c = 2\cdot22 \, \frac{\text{c.c.}}{\text{gr.}}$, than those obtained from van der Waals' equation.

§ 31. Regarding the transition from the liquid to the solid state, the same considerations hold as for that from the gaseous to the liquid state. In this case also one can draw the system of isotherms, and establish " theoretical " regions of the isotherms. In fact there are certain phenomena, as, for example, the supercooling of liquids, which point to a more or less unstable liquid state.

In its most complete form the characteristic equation would comprise the gaseous, liquid, and solid states simultaneously. No formula of such generality, however, has as yet been established for any substance.

§ 32. Mixtures.—While, as shown in § 19, the characteristic equation of a mixture of perfect gases reduces in a simple manner to that of its components, no such simplification takes place, in general, when substances of any kind are mixed. Only for gases and vapours does Dalton's law hold, at least with great approximation, that the total pressure of a mixture is the sum of the partial pressures which each gas would exert if it alone filled the total volume at the given temperature. This law enables us to establish the characteristic equation of any gas mixture, provided that of the constituent gases be known. It also decides the ques-

---

From (2) and (3), $v = 3a + 2b$ . . . . (4)

Substituting (4) in (2) and reducing, we get

$$T^2 = \frac{8c}{27(a+b)R} \quad \cdot \quad \cdot \quad \cdot \quad \cdot \quad \cdot \quad \text{(5)}$$

And substituting (4) and (5) in (1) and reducing, we have

$$p^2 = \frac{cR}{216(a+b)^3} \quad \cdot \quad \cdot \quad \cdot \quad \cdot \quad \cdot \quad \text{(6)—Tr.}$$

tion, unanswered in § 17, whether to the individual gases of a mixture common pressure and different volumes, or common volume and different pressures, should be ascribed. From the consideration of a vapour differing widely from an ideal gas, it follows that the latter of these views is the only one admissible. Take, for instance, atmospheric air and water vapour at 0° C. under atmospheric pressure. Here the water vapour cannot be supposed to be subject to a pressure of 1 atm., since at 0° C. no water vapour exists at this pressure. The only choice remaining is to assign to the air and water vapour a common volume (that of the mixture) and different pressures (partial pressures).

For mixtures of solid and liquid substances no law of general validity has been found, that reduces the characteristic equation of the mixture to those of its constituents.

CHAPTER II.

MOLECULAR WEIGHT.

§ **33.** In the preceding chapter only such physical changes have been discussed as concern temperature, pressure, and density. The chemical constitution of the substance or mixture in question has been left untouched. Cases are frequent, however (much more so, in fact, than was formerly supposed), in which the chemical nature of a substance is altered by a change of temperature or pressure. The more recent development of thermodynamics has clearly brought out the necessity of establishing a fundamental difference between physical and chemical changes such as will exclude continuous transition from the one kind to the other (*cf.* § 42, *et seq.*, and § 238). It has, however, as yet not been possible to establish a practical criterion for distinguishing them, applicable to all cases. However strikingly most chemical processes differ from physical ones in their violence, suddenness, development of heat, changes of colour and other properties, yet there are, on the other hand, numerous changes of a chemical nature that take place with continuity and comparative slowness; for example, dissociation. One of the main tasks of physical chemistry in the near future will be the further elucidation of this essential difference.

§ **34.** The peculiar characteristic of all chemical reactions is that they take place according to constant proportions by weight. A certain weight (strictly speaking, a mass) may therefore be used as a characteristic expression for the nature of the reaction of a given chemically homogeneous substance, whether an element or a compound. Such a weight is called an *equivalent weight*. It is arbitrarily fixed for one element— *e.g.* for hydrogen at 1 gr.—and then the equivalent weight of any other element (*e.g.* oxygen) is that weight which will combine with 1 gr. of hydrogen. The weight of the compound thus formed is, at the same time, its equivalent weight. By proceeding in this way, the equivalent weights of all chemically homogeneous substances may be found. The equivalent weights of elements that do not combine directly with hydrogen can easily be determined, since in every case a number of elements can be found that combine directly with the element in question and also with hydrogen.

The total weight of a body divided by its equivalent weight is called the *number of equivalents* contained in the body. Hence we may say that, in every chemical reaction, an equal number of equivalents of the different substances react with one another.

§ 35. There is, however, some ambiguity in the above definition, since two elements frequently combine in more ways than one. For such cases there would exist several values of the equivalent weight. Experience shows, however, that the various possible values are always simple multiples or submultiples of any one of them. The ambiguity in the equivalent weight, therefore, reduces itself to multiplying or dividing that quantity by a simple integer. We must accordingly generalize the foregoing statement, that an equal number of equivalents react with one another, and say, that the number of equivalents that react with one another are in simple numerical proportions.* Thus 16 parts by weight of oxygen combine with 28 parts by weight of nitrogen to form nitrous oxide, or with 14 parts to form nitric oxide, or with $9\frac{1}{3}$ parts to form nitrous anhydride, or

* If these whole numbers could be of any magnitude then the equivalent weight would be a continuous varying quantity. Each quantity could then, to any approximation, be represented by the ratio of two whole numbers. The value of the equivalent weights could not be defined from the weights reacting with one another, and the whole above considerations would be illusory. The discontinuity in the variation of equivalent weights is characteristic of the chemical nature of a substance, in contrast to its physical properties. To sum up shortly, one may say that physical changes take place continuously, while chemical changes take place discontinuously. Physics deals chiefly with continuous varying quantities, while chemistry deals chiefly with whole numbers.

with 7 parts to form nitrogen tetroxide, or with $5\frac{3}{8}$ parts to form nitric anhydride. Any one of these numbers may be assigned to nitrogen as its equivalent weight, if 16 be taken as that of oxygen. They are, however, in simple rational proportions, since

$$28 : 14 : 9\tfrac{1}{3} : 7 : 5\tfrac{3}{5} = 60 : 30 : 20 : 15 : 12.$$

§ **36.** The ambiguity in the definition of the equivalent weight of nitrogen, exemplified by the above series of numbers, is removed by selecting a particular one of them to denote the *molecular weight* of nitrogen. In the definition of the molecular weight as a quite definite quantity depending only on the particular state of a substance, and independent of possible chemical reactions with other substances, lies one of the most important and most fruitful achievements of theoretical chemistry. Its exact statement can at present be given only for special cases, viz. for perfect gases and dilute solutions. We need consider only the former of these, as we shall see from thermodynamics that the latter is also thereby determined.

The definition of the molecular weight for a chemically homogeneous perfect gas is rendered possible by the further empirical law, that gases combine, not only in simple multiples of their equivalents, but also, at the same temperature and pressure, in simple volume proportions (Gay Lussac). It immediately follows that the number of equivalents, contained in equal volumes of different gases, must bear simple ratios to one another. The values of these ratios, however, are subject to the above-mentioned ambiguity in the selection of the equivalent weight. The ambiguity is, however, removed by putting all these ratios = 1, *i.e.* by establishing the condition that equal volumes of different gases shall contain an equal number of equivalents. Thus a definite choice is made from the different possible values, and a definite equivalent weight obtained for the gas, which is henceforth denoted as the *molecular weight* of the gas. At the same time the number of equivalents in a quantity of the gas, which may be found by dividing the total weight by the molecular weight, is defined as the *number* * *of molecules* contained in that quantity. Hence, *equal volumes of perfect gases at the same temperature and pressure contain an equal number of molecules* (Avogadro's law). The molecular weights of chemically homogeneous gases are, therefore, directly proportional to the masses contained in equal volumes, *i.e.* to the densities. The ratio of the densities is equal to the ratio of the molecular weights. If we call m_1 and m_2 the molecular weights of two perfect gases, then according to § 11

$$m_1 : m_2 = \frac{1}{C_1} : \frac{1}{C_2} \quad . \quad . \quad . \quad . \quad (12b)$$

§ 37. Putting the molecular weight of hydrogen $= m_H$, that of any other chemically homogeneous gas must be equal to m_H multiplied by its specific density relative to hydrogen (§ 11). The following table gives the specific densities of some gases and vapours, relative to hydrogen, in round numbers, and also their molecular weights :

	Specific Density.	Molecular Weight.
Hydrogen	1	m_H
Oxygen	16	16 m_H
Nitrogen	14	14 m_H
Water vapour . . .	9	9 m_H
Ammonia	8·5	8·5 m_H
Nitrous oxide	22	22 m_H
Nitric oxide	15	15 m_H

With the help of this table we can completely answer the question as to how the molecular weight of a compound is built up out of the molecular weights of its elements.

Water vapour consists of 1 part by weight of hydrogen and 8 parts by weight of oxygen, therefore the molecule of water vapour, 9 m_H, must consist of m_H parts by weight of hydrogen and 8 m_H parts by weight of oxygen—*i.e.* according to the above table, of one molecule of hydrogen and half a

---

* This, of course, need not be a whole number, since we are concerned here, not with the real molecules in the sense of the atomic theory, but with the gr.-molecules, or *mols.*, as arbitrarily defined.

molecule of oxygen. Further, since ammonia, according to analysis, consisting of 1 part by weight of hydrogen and $4\frac{2}{3}$ parts by weight of nitrogen, its molecule $8 \cdot 5\ m_H$ must necessarily contain $1 \cdot 5\ m_H$ parts by weight of hydrogen and $7\ m_H$ parts by weight of nitrogen—*i.e.* according to the table, $1\frac{1}{2}$ molecules of hydrogen and $\frac{1}{2}$ molecule of nitrogen. Further, since nitrous oxide, according to analysis, contains 16 parts by weight of oxygen and 28 parts by weight of nitrogen, the molecule of nitrous oxide, $22\ m_H$, necessarily contains $8\ m_H$ parts by weight of oxygen and $14\ m_H$ parts by weight of nitrogen, *i.e.* according to the table, $\frac{1}{2}$ molecule of oxygen and 1 molecule of nitrogen. Thus Avogadro's law enables us to give in quite definite numbers the molecular quantities of each constituent present in the molecule of any chemically homogeneous gas, provided we know its density and its chemical composition.

§ **38.** The smallest weight of a chemical element entering into the molecules of its compounds is called an *atom*. Hence half a molecule of hydrogen is called an atom of hydrogen, H; similarly, half a molecule of oxygen an atom of oxygen, O; and half a molecule of nitrogen an atom of nitrogen, N. The diatomic molecules of these substances are represented by H_2, O_2, N_2. An atom of mercury, on the contrary, is equal to a whole molecule, because in the molecules of its compounds only whole molecules of mercury vapour occur. In order to determine numerical values for the atomic and molecular weights, it is still necessary to arbitrarily fix the atomic weight of some element. Formerly it was usual to put $H = 1$ and $O = 16$. It has, however, been shown that the ratio of the equivalent weights of oxygen and hydrogen is not exactly 16, but about $15 \cdot 87$. Since the oxygen compounds of most elements can be more accurately analysed than the hydrogen compounds, it has become customary to start off from the atomic weight of oxygen, $O = 16$, as a definition. Then the molecular weight of oxygen,

$$O_2 = 32 = 15 \cdot 87\ m_H.$$

It follows that the molecular weight of hydrogen,

$$m_H = \frac{32}{15 \cdot 87} = 2 \cdot 016 = H_2,$$

and the atomic weight of hydrogen,

$$H = 1 \cdot 008.$$

The molecular weights of the above table then become :

Molecular Weight.

Hydrogen	$2 \cdot 016 = H_2$
Oxygen	$32 \cdot 00 = O_2$
Nitrogen	$28 \cdot 02 = N_2$
Water vapour	$18 \cdot 02 = H_2O$
Ammonia	$17 \cdot 03 = NH_3$
Nitrous oxide . . . . . ., . .	$44 \cdot 02 = N_2O$
Nitric oxide	$30 \cdot 01 = NO$

§ 39. In general, then, the molecular weight of a chemically homogeneous gas is 2·016 times its specific density relative to hydrogen, or equal to 32 times its specific density relative to oxygen. Conversely, if the molecular weight, m, of a gas be known, its specific density, and consequently the constant C in the characteristic equation (5), can be calculated. If we denote all quantities referring to oxygen by the suffix 0, then, according to equation (12b) in § 36,

$$C = \frac{m_0 C_0}{m} \quad . \quad . \quad . \quad . \quad . \quad (13)$$

Now $m_0 = 32$, and the constant C_0 can be calculated from the density of oxygen at 0° C. and under atmospheric pressure. According to the table in § 11,

$$\frac{1}{v_0} = 0 \cdot 0014291$$
$$p = 1013250 \ (\S \ 7)$$
$$T = 273$$

Therefore, $$C_0 = \frac{p v_0}{T}$$

and, by (13), $$C = \frac{m_0}{m} \frac{p v_0}{T},$$

or, $$C = \frac{32 \times 1013250}{m \times 273 \times 0 \cdot 0014291} = \frac{83110000}{m} = \frac{m_0 C_0}{m}.$$

If we had started off from the molecular weight of hydrogen, we would have obtained a value not materially greater.

If, for shortness, we put

$$8 \cdot 315 \times 10^7 = R \quad . \quad . \quad . \quad . \quad (13a)$$

then the general equation of a perfect chemically homogeneous gas of molecular weight m becomes

$$p = \frac{R}{m} \cdot \frac{T}{v} \quad . \quad . \quad . \quad . \quad . \quad (14)$$

where R, being independent of the nature of the individual gas, is generally called the absolute gas constant. The molecular weight may be deduced directly from the characteristic equation by the aid of the constant R, since

$$m = \frac{R}{C} \quad . \quad . \quad . \quad . \quad . \quad (15)$$

Since $v = \dfrac{V}{M}$, we have

$$V = \frac{RT}{p} \cdot \frac{M}{m}.$$

But $\dfrac{M}{m}$ is the quantity defined above as the number of molecules in the gas, and, therefore, if $\dfrac{M}{m} = n$,

$$V = \frac{RT}{p} \cdot n,$$

which means that at a given temperature and pressure the volume of a quantity of gas depends only on the number of the molecules present, and not at all on the nature of the gas, as Avogadro's law requires.

§ **40.** In a mixture of chemically homogeneous gases of molecular weights m_1, m_2 . . . the relation between the partial pressures is, according to (9),

$$p_1 : p_2 : \ldots = C_1 M_1 : C_2 M_2 \ldots$$

But in (15) we have $\quad C_1 = \dfrac{R}{m_1}; \; C_2 = \dfrac{R}{m_2}; \; \ldots$

$$\therefore \; p_1 : p_2 : \ldots = \frac{M_1}{m_1} : \frac{M_2}{m_2} : \ldots = n_1 : n_2 \ldots$$

i.e. the ratio of the partial pressures is also the ratio of the number of molecules of each gas present. Equation (10) gives for the total volume

$$V = \frac{(C_1 M_1 + C_2 M_2 + \ldots)T}{p}$$
$$= \frac{RT}{p}\left(\frac{M_1}{m_1} + \frac{M_2}{m_2} + \ldots\right)$$
$$= \frac{RT}{p}(n_1 + n_2 + \ldots)$$
$$= \frac{RT}{p}n \quad . \quad . \quad . \quad . \quad . \quad . \quad (16)$$

The volume of the mixture is therefore determined by the total number of the molecules present, just as in the case of a chemically homogeneous gas.

§ 41. It is evident that we cannot speak of the molecular weight of a mixture. Its *apparent* or *mean* molecular weight, however, may be defined as the molecular weight which a chemically homogeneous gas would have if it contained in the same mass the same number of molecules as the mixture. If we denote the apparent molecular weight by m, we have

$$\frac{M_1 + M_2 + \ldots}{m} = \frac{M_1}{m_1} + \frac{M_2}{m_2} + \ldots$$

and

$$n = \frac{M_1 + M_2 + \ldots}{\dfrac{M_1}{m_1} + \dfrac{M_2}{m_2} + \ldots}$$

The apparent molecular weight of air may thus be calculated. Since

$$m_1 = O_2 = 32; \quad m_2 = N_2 = 28; \quad M_1 : M_2 = 0.30 \ (\S 20)$$

we have

$$m = \frac{0.30 + 1}{\dfrac{0.30}{32} + \dfrac{1}{28}} = 28.9,$$

which is somewhat larger than the molecular weight of nitrogen.

§ **42.** The characteristic equation of a perfect gas, whether chemically homogeneous or not, gives, according to (16), the total number of molecules, but yields no means of deciding, as has been pointed out in § 19, whether or not these molecules are all of the same kind. In order to answer this question, other methods must be resorted to, none of which, however, is practically applicable to all cases. A decision is often reached by an observation of the process of diffusion through a porous or, better, a semi-permeable membrane. The individual gases of a mixture will separate from each other by virtue of the differences in their velocities of diffusion, which may even sink to zero in the case of semi-permeable membranes, and thus disclose the inhomogeneity of the substance. The chemical constitution of a gas may often be inferred from the manner in which it originated. It is by means of the expression for the entropy (§ 237) that we first arrive at a fundamental definition for a chemically homogeneous gas.

§ **43.** Should a gas or vapour not obey the laws of perfect gases, or, in other words, should its specific density depend on the temperature or the pressure, Avogadro's definition of molecular weight is nevertheless applicable. By equation (16) we have $n = \dfrac{p\mathrm{V}}{\mathrm{R}T}$. The number of molecules in this case, instead of being a constant, will be dependent upon the momentary physical condition of the substance. We may, in such cases, either assume the number of molecules to be variable, or refrain from applying Avogadro's definition of the number of molecules. In other words, the cause for the deviation from the ideal state may be sought for either in the chemical or physical conditions. The latter view preserves the chemical nature of the gas. The molecules remain intact under changes of temperature and pressure, but the characteristic equation is more complicated than that of Boyle and Gay Lussac—like that, for example, of van der Waals or of Clausius. The other view differs essentially

from this, in that it represents any gas, not obeying the laws
of perfect gases, as a mixture of various kinds of molecules
(in nitrogen peroxide N_2O_4 and NO_2, in phosphorus penta-
chloride PCl_5, PCl_3, and Cl_2). The volume of these is supposed
to have at every moment the exact value theoretically required
for the total number of molecules of the mixture of these gases.
The volume, however, does not vary with temperature and
pressure in the same way as that of a perfect gas, because
chemical reactions take place between the different kinds of
molecules, continuously altering the number of each kind
present, and thereby also the total number of molecules in
the mixture. This hypothesis has proved fruitful in cases of
great differences of density—so-called abnormal vapour
densities—especially where, beyond a certain range of temperature or pressure, the specific density once more becomes
constant. When this is the case, the chemical reaction has
been completed, and for this reason the molecules henceforth
remain unchanged. Amylene hydrobromide, for instance,
acts like a perfect gas below 160° and above 360°, but shows
only half its former density at the latter temperature. The
doubling of the number of molecules corresponds to the
equation

$$C_5H_{11}Br = C_5H_{10} + HBr.$$

Mere insignificant deviations from the laws of perfect gases
are generally attributed to physical causes—as, *e.g.*, in water
vapour and carbon dioxide—and are regarded as the fore-
runners of condensation. The separation of chemical from
physical actions by a principle which would lead to a more
perfect definition of molecular weight for variable vapour
densities, cannot be accomplished at the present time. The
increase in the specific density which many vapours exhibit
near their point of condensation might just as well be attributed to such chemical phenomena as the formation of double
or multiple molecules.* In fact, differences of opinion exist

---

* W. Nernst ("Verhandlungen der deutschen Phys. Ges.," II, S. 313, 1909)
has made the characteristic equation of water vapour depend upon the
formation of double molecules $(H_2O)_2$.

in a number of such cases. The molecular weight of sulphur vapour below 800°, for instance, is generally assumed to be $S_6 = 192$; but some assume a mixture of molecules $S_8 = 256$ and $S_2 = 64$, and others still different mixtures. In doubtful cases it is safest, in general, to leave this question open, and to admit both chemical and physical changes as causes for the deviations from the laws of perfect gases. This much, however, may be affirmed, that for small densities the physical influences will be of far less moment than the chemical ones, for, according to experience, all gases approach the ideal condition as their densities decrease (§ 21). This is an important point, which we will make use of later.

CHAPTER III.

§ **44.** IF we plunge a piece of iron and a piece of lead, both of equal weight and at the same temperature (100° C.), into two precisely similar vessels containing equal quantities of water at 0° C., we find that, after thermal equilibrium has been established in each case, the vessel containing the iron has increased in temperature much more than that containing the lead. Conversely, a quantity of water at 100° is cooled to a much lower temperature by a piece of iron at 0°, than by an equal weight of lead at the same temperature. This phenomenon leads to a distinction between *temperature* and *quantity of heat.* As a measure of the heat given out or received by a body, we take the increase or decrease of temperature which some *normal* substance (*e.g.* water) undergoes when it alone is in contact with the body, provided all other causes of change of temperature (as compression, etc.) are excluded. The quantity of heat given out by the body is assumed to be equal to that received by the normal substance, and *vice versâ.* The experiment described above proves, then, that a piece of iron in cooling through a given interval of temperature gives out more heat than an equal weight of lead (about four times as much), and conversely, that, in order to bring about a certain increase of temperature, iron requires a correspondingly larger supply of heat than lead.

§ **45.** It was, in general, customary to take as the unit of heat that quantity which must be added to 1 gr. of water to raise its temperature from 0° C. to 1° C. (zero calorie). This is almost equal to the quantity of heat which will raise 1 gr. of water 1° C. at any temperature. The refinement of calorimetric measurements has since made it necessary to take account of the initial temperature of the water, and it is often found convenient to define the calorie as that quantity of heat which will raise 1 gr. of water from 14·5° C. to 15·5° C. This is about $\dfrac{1}{1\cdot008}$ of a zero calorie. Finally, a *mean calorie* has been introduced, namely, the hundredth part of the heat required to raise 1 gr. of water from 0° C. to 100° C. The mean calorie is about equal to the zero calorie. Besides these so-called *small* calories, there are a corresponding number of *large* or kilogram calories, which contain 1000 small calories.

§ **46.** The ratio of the quantity of heat Q absorbed by a body to the corresponding rise of temperature $T' - T = \varDelta T$ is called the mean *heat capacity* of the body between the temperatures T and T' :

$$\frac{Q}{\varDelta T} = C_m.$$

The heat capacity per grm. is called the *specific heat* of the substance :

$$c_m = \frac{C_m}{M} = \frac{Q}{M \cdot \varDelta T} = \frac{q}{\varDelta T}.$$

Hence the mean specific heat of water between 0° C. and 1° C. is equal to the zero calorie. If we take infinitely small intervals of temperature, we obtain, for the heat capacity of a body, and for the specific heat at temperature T,

$$\frac{Q}{dT} = C \text{ and } \frac{q}{dT} = c.$$

The specific heat, in general, varies with the temperature, but, mostly, very slowly. It is usually permissible to put the specific heat at a certain temperature equal to the mean specific heat of an adjoining interval of moderate size.

§ **47.** Heat capacity and specific heat require, however, more exact definition. Since the thermodynamic state of a body depends not only on the temperature but also on a second variable, for example the pressure, changes of state, which accompany changes of temperature, are not determined, unless the behaviour of the second variable is also given. Now it is true that the heat capacity of solids and liquids is nearly independent of any variations of external pressure that may take place during the process of heating. Hence the definition of the heat capacity is not, usually, encumbered with a condition regarding pressure. The specific heat of gases, however, is influenced considerably by the conditions of the heating process. In this case the definition of specific heat would, therefore, be incomplete without some statement as to the accompanying conditions. Nevertheless, we speak of the specific heat of a gas, without further specification, when we mean its specific heat at constant (atmospheric) pressure, as this is the value most readily determined.

§ **48.** That the heat capacities of different substances should be referred to unit mass is quite arbitrary. It arises from the fact that quantities of matter can be most easily compared by weighing them. Heat capacity might, quite as well, be referred to unit volume. It is more rational to compare masses which are proportional to the molecular and atomic weights of substances, for then certain regularities at once become manifest. The corresponding heat capacities are obtained by multiplying the specific heats by the molecular or atomic weights. The values thus obtained are known as the *molecular* or *atomic heats*.

§ **49.** The chemical elements, especially those of high atomic weight, are found to have nearly the constant atomic heat of 6·3 (Dulong and Petit). It cannot be claimed that this law is rigorously true, since the heat capacity depends on the molecular constitution, as in the case of carbon, as well as on the temperature. The effect of temperature is especially marked in the elements, carbon, boron, and silicon which show the largest deviations from Dulong and Petit's law. The conclusion is, however, justified, that Dulong and Petit's law is founded on some more general law of nature. The derivation of this law would, however, take us beyond the region of general thermodynamics.

§ **50.** Similar regularities, as appear in the atomic heats of elements, are also found in the molecular heats of compounds, especially with compounds of similar chemical constitution. According to F. Neumann's law, subsequently confirmed by Regnault and particularly by von Kopp, the molecular heat of a solid compound is simply equal to the sum of the atomic heats of the elements contained in it, each element maintaining its particular atomic heat, whether it be 6·3, according to Dulong and Petit's law, or not. This relation also is only approximately true.

§ **51.** Since all calorimetric measurements, according to § 44, extend only to quantities of heat imparted to bodies or given out by them, they do not lead to any conclusion as to the total amount of heat *contained in* a body of given temperature. It would be absurd to define the heat contained in a body of given temperature, density, etc., as the number of calories absorbed by the body in its passage from some normal state into its present state, for the quantity thus defined would assume different values according to the way in which the change was effected. A gas at 0° and atmospheric pressure can be brought to a state where its temperature is 100° and its pressure 10 atmospheres, either by heating to 100° under constant pressure, and then compressing at constant temperature; or by compressing isothermally to 10 atmospheres, and then heating isobarically to 100°; or, finally, by compressing and heating simultaneously or alternately in a variety of ways. The total number of calories absorbed would in each case be different (§ 77). It is seen, then, that it is useless to speak of a certain quantity of heat which must be applied to a body in a given state to bring it to some other state. If the " total heat contained in a body " is to be expressed numerically, as is done in the kinetic theory of heat, where the heat of a body is defined as the total energy of its internal motions, it must not be interpreted as the sum-total of the quantities of heat applied to the body. As we shall make no use of this quantity in our present work, no definition of it need be attempted.

§ 52. In contrast to the above representation of the facts, the older (Carnot's) theory of heat, which started from the hypothesis that heat is an indestructible substance, necessarily reached the conclusion that the " heat contained in a body " depends solely on the number of calories absorbed or given out by it. The heating of a body by other means than direct application of heat, by compression or by friction for instance, according to that theory produces no change in the " total heat." To explain the rise of temperature which takes place notwithstanding, it was necessary to make the assumption that compression and friction so diminish the body's heat capacity, that the same amount of heat now produces a higher temperature, just as, for example, a moist sponge appears more moist if compressed, although the quantity of liquid in the sponge remains the same. In the meantime, Rumford and Davy proved by direct experiment that bodies, in which any amount of heat can be generated by an adequate expenditure of work, do not in the least alter their heat capacities with friction. Regnault, likewise, showed, by accurate measurements, that the heat capacity of gases is independent of or only very slightly dependent on volume; that it cannot, therefore, diminish, in consequence of compression, as much as Carnot's theory would require. Finally, W. Thomson and Joule have demonstrated by careful experiments that a gas, when expanding without overcoming external pressure, undergoes no change of temperature, or an exceedingly small one (*cf.* § 70), so that the cooling of gases generally observed when they expand is not due to the increase of volume *per se*, but to the work done in the expansion. Each one of these experimental results would by itself be sufficient to disprove the hypothesis of the indestructibility of heat, and to overthrow the older theory.

§ **53.** While, in general, the heat capacity varies continuously with temperature, every substance possesses, under certain external pressures, so-called *singular* values of temperature, for which the heat capacity, together with other properties, is discontinuous. At such temperatures the heat absorbed no longer affects the entire body, but only one of the parts into which it has split; and it no longer serves to increase the temperature, but simply to alter the state of aggregation, *i.e.* to melt, evaporate, or sublime. Only when the entire substance has again become homogeneous will the heat imparted produce a rise in temperature, and then the heat capacity becomes once more capable of definition. The quantity of heat necessary to change 1 gram of a substance from one state of aggregation to another is called the *latent heat*, in particular, the *heat of fusion, of vaporization,* or *of sublimation.* The same amount of heat is set free when the substance returns to its former state of aggregation. Latent heat, as in the case of specific heat, is best referred, not to unit mass, but to molecular or atomic weight. Its amount largely depends on the external conditions under which the process is carried out (§ 47), constant pressure being the most important condition.

§ **54.** Like the changes of the state of aggregation, all processes involving mixture, or solution, and all chemical reactions are accompanied by a development of heat of greater or less amount, which varies according to the external conditions. This we shall henceforth name the *heat of reaction* of the process under consideration, in particular as the heat of mixture, of solution, of combination, of dissociation, etc. It is reckoned *positive* when heat is set free or developed, *i.e.* given out by the body (exothermal processes); *negative*, when heat is absorbed, or rendered latent, *i.e.* taken up by the body (endothermal processes).

PART II.

THE FIRST FUNDAMENTAL PRINCIPLE OF THERMODYNAMICS.

CHAPTER I.

GENERAL EXPOSITION.

§ 55. THE *first law* of thermodynamics is nothing more than the principle of the conservation of energy applied to phenomena involving the production or absorption of heat. Two ways lead to a deductive proof of this principle. We may take for granted the correctness of the mechanical view of nature, and assume that all changes in nature can be reduced to motions of material points between which there act forces which have a potential. Then the principle of energy is simply the well-known mechanical theorem of kinetic energy, generalized to include all natural processes. Or we may, as is done in this work, leave open the question concerning the possibility of reducing all natural processes to those of motion, and start from the fact which has been tested by centuries of human experience, and repeatedly verified, viz. that *it is in no way possible, either by mechanical, thermal, chemical, or other devices, to obtain perpetual motion, i.e.* it is impossible to construct an engine which will work in a cycle and produce continuous work, or kinetic energy, from nothing. We shall not attempt to show how this single fact of experience, quite independent of the mechanical view of nature, serves to prove the principle of energy in its generality, mainly for the reason that the validity of the energy principle is nowadays no longer disputed. It will be different, however, in the case of the *second law* of thermodynamics, the proof of which, at the present stage of the development of our subject, cannot be too carefully presented. The general validity of this law is still contested from time to time, and its significance variously interpreted, even by the adherents of the principle.

§ **56. The energy** of a body, or system of bodies, is a magnitude depending on the momentary condition of the system. In order to arrive at a definite numerical expression for the energy of the system in a given state, it is necessary to fix upon a certain *normal* arbitrarily selected state (*e.g.* 0° C. and atmospheric pressure). The energy of the system in a given state, referred to the arbitrarily selected normal state, is then equal to *the algebraic sum of the mechanical equivalents of all the effects produced outside the system when it passes in any way from the given to the normal state.* The energy of a system is, therefore, sometimes briefly denoted as the faculty to produce external effects. Whether or not the energy of a system assumes different values according as the transition from the given to the normal state is accomplished in different ways is not implied in the above definition. It will be necessary, however, for the sake of completeness, to explain the term "mechanical equivalent of an external effect."

§ **57.** Should the external effect be mechanical in nature —should it consist, *e.g.*, in lifting a weight which loads the system, overcoming atmospheric pressure, or producing kinetic energy—then its mechanical equivalent is simply equal to the mechanical work done by the system on the external body (weight, atmosphere, projectile). It is positive if the displacement take place in the direction of the force exercised by the system—when the weight is lifted, the atmosphere pushed back, the projectile discharged,—negative in the opposite sense.

But if the external effect be thermal in nature—if it consist, *e.g.*, in heating surrounding bodies (the atmosphere, a calorimetric liquid, etc.)—then its mechanical equivalent is equal to the number of calories which will produce the same rise of temperature in the surrounding bodies multiplied by an absolute constant, which depends only on the units of heat and mechanical work, the so-called *mechanical equivalent of heat*. This proposition, which appears here only as a definition, receives through the principle of the conservation of energy a physical meaning, which may be put to experimental test.

§ 58. The Principle of the Conservation of Energy asserts, generally and exclusively, that the energy of a system in a given state, referred to a fixed normal state, has a quite definite value; in other words—substituting the definition given in § 56—that *the algebraic sum of the mechanical equivalents of the external effects produced outside the system, when it passes from the given to the normal state, is independent of the manner of the transformation.* On passing into the normal state the system thus produces a definite total of effects, as measured in mechanical units, and it is this sum—the " work-value " of the external effects —that represents the energy of the system in the given state.

§ 59. The validity of the principle of the conservation of energy may be experimentally verified by transferring a system in various ways from a given state to a certain other state, which may here be called the normal state, and then testing whether the mechanical equivalents of the external effects, calculated as laid down in § 57, give the same sum in each case. Special care must be taken that the initial and final states of the system are the same each time and that none of the external effects is overlooked or taken into account more than once.

§ 60. As a first application we shall discuss Joule's famous experiments, in which the external effects produced by weights falling from a certain height were compared, first, when performing only mechanical work (*e.g.* lifting a load), and second, when by suitable contrivances generating heat by friction. The initial and final position of the weights may be taken as the two states of the system, the work or heat produced, as the external effects. The first case, where the weights produce only mechanical work, is simple, and requires no experiment. Its mechanical equivalent is the product of the sum of the weights, and the height through which they fall, quite independent of the nature of the apparatus used to perform the mechanical work (lever, pulley, etc.) and of the load raised. This independence is a condition of the principle of the conservation of energy (§ 58). The second case requires accurate measurement of the increase of temperature, which the surrounding substances (water, mercury) undergo in consequence of the friction, as well as of their heat capacities, for the determination of the number of calories which will produce in them the same rise of temperature. It is, of course, entirely immaterial what our views may be with regard to the details of the frictional generation of heat, or with regard to the ultimate form of the heat thus generated. The only point of importance is that the state produced in the liquid by friction is identical with a state produced by the absorption of a definite number of calories.

Joule, according to the principle of the conservation of energy, equated the mechanical work, corresponding to the fall of the weights, to the mechanical equivalent of the heat produced by friction, and found that the mechanical equivalent of a gram-calorie is, under all circumstances, equal to the work done in lifting a weight of a gram through a height of 423·55 meters. That all his experiments with different weights, different calorimetric substances, and different temperatures, led to the same value, goes to prove the correctness of the principle of the conservation of energy.

§ **61.** In order to determine the mechanical equivalent of heat in absolute units, we must bear in mind that Joule's measurements were made by a mercury thermometer and expressed in laboratory calories (§ 45). At room temperature, 1° on Joule's mercury thermometer represents about $\frac{1}{1\cdot007}$ of 1° on the gas thermometer. A calorie on the gas thermometer scale has, therefore, a mechanical equivalent of $423\cdot55 \times 1\cdot007 = 427$ grm. meters.

The acceleration of gravity must also be considered, since raising a gram to a certain height represents, in general, different amounts of work in different latitudes. The absolute value of the work done is obtained by multiplying the weight, *i.e.* the product of the mass and the acceleration of gravity, by the height of fall. The following table gives the mechanical equivalent of heat in the different calories :

Unit of heat referred to the gas thermometer.	Corresponding height in meters to which 1 gr. must be raised in places of mean latitude.	Absolute value of the mechanical equivalent in the c.g.s. system (erg).
15° calorie	427	$4\cdot19 \times 10^7$
Zero calorie	430	$4\cdot22 \times 10^7$

The numbers of the last column are derived from those of the preceding one by multiplying by 98,100, to reduce grams to dynes, and meters to centimeters. Joule's results have been substantially confirmed by recent careful measurements by Rowland and others.

§ 62. The determination of the mechanical equivalent of heat enables us to express quantities of heat in ergs directly, instead of in calories. The advantage of this is, that a quantity of heat is not only proportional to, but directly equal to its mechanical equivalent, whereby the mathematical expression for the energy is greatly simplified. This unit of heat will be used in all subsequent equations. The return to calories is, at any time, readily accomplished by dividing by $4\cdot19 \times 10^7$.

§ 63. Some further propositions immediately follow from the above exposition of the principle of energy. The energy, as stated, depends on the momentary condition of the system. To find the change of energy, $U_1 - U_2$, accompanying the transition of the system from a state 1 to a state 2, we should, according to the definition of the energy in § 58, have to measure U_1 as well as U_2 by the mechanical equivalent of the external effects produced in passing from the given states to the normal state. But, supposing we so arrange matters that the system passes from state 1, through state 2, into the normal state, it is evident then that $U_1 - U_2$ is simply the mechanical equivalent of the external effects produced in passing from 1 to 2. The decrease of the energy of a system subjected to any change is, then, the mechanical equivalent of the external effects resulting from that change; or, in other words, the *increase of the energy* of a system which undergoes any change, is equal to the mechanical equivalent of the heat absorbed and the work expended in producing the change :

$$U_2 - U_1 = Q + W \quad . \quad . \quad . \quad . \quad (17)$$

Q is the mechanical equivalent of the heat absorbed by the system, *e.g.* by conduction, and W is the amount of work expended on the system. W is positive if the change takes place in the direction of the external forces. The sum $Q + W$ represents the mechanical equivalent of all the thermal and mechanical operations of the surrounding bodies on the system. We shall use Q and W always in this sense.

The value of $Q + W$ is independent of the manner of the transition from 1 to 2, and evidently also of the selection of the normal state. When differences of energy of one and the same system are considered, it is, therefore, not even necessary to fix upon a normal state. In the expression for the energy of the system there remains then an arbitrary additive constant undetermined.

§ **64.** The difference $U_2 - U_1$ may also be regarded as the energy of the system in state 2, referred to state 1 as the normal state. For, if the latter be thus selected, then

$U_1 = 0$, since it takes no energy to change the system from 1 to the normal state, and $U_2 - U_1 = U_2$. The normal state is, therefore, sometimes called the zero state.

§ 65. States 1 and 2 may be identical, in which case the system changing from 1 to 2 passes through a so-called *cycle of operations.* In this case,

$$U_2 = U_1 \text{ and } Q + W = 0 \quad . \quad . \quad . \quad (18)$$

The mechanical equivalent of the external effects is zero, or the external heat effect is equal in magnitude and opposite in sign to the external work. This proposition shows the impracticability of perpetual motion, which necessarily presupposes engines working in complete cycles.

§ 66. If no external effects ($Q = 0$, $W = 0$) be produced by a change of state of the system, its energy remains constant (conservation of the energy). The quantities, on which the state of the system depends, may undergo considerable changes in this case, but they must obey the condition $U = $ const.

A system which changes without being acted on by external agents is called a *perfect* or *isolated system.* Strictly speaking, no perfect system can be found in nature, since there is constant interaction between all material bodies of the universe, and the law of the *conservation* of energy cannot be rigorously applied to any real system. It is, however, of importance to observe that by an adequate choice of the system which is to undergo the contemplated change, we have it in our power to make the external effect as small as we please, in comparison with the changes of energy of portions of the system itself. Any particular external effect may be eliminated by making the body which produces this effect, as well as the recipient, a part of the system under consideration. In the case of a gas which is being compressed by a weight sinking to a lower level, if the gas by itself be the system considered, the external effect on it is equal to the work done by the weight. The energy of the system accordingly increases. If, however, the weight and the earth be considered parts of the system, all external effects are eliminated, and the energy of this system remains constant. The expression for the energy now contains a new term representing the potential energy of the weight. The loss of the potential energy of the weight is exactly compensated by the gain of the internal energy of the gas. All other cases admit of similar treatment.

CHAPTER II.

APPLICATIONS TO HOMOGENEOUS SYSTEMS

§ **67.** WE shall now apply the first law of thermodynamics as expressed in equation (17),

$$U_2 - U_1 = Q + W,$$

to a homogeneous substance, whose state is determined, besides by its chemical nature and mass M, by two variables, the temperature T and the volume V, for instance. The term *homogeneous* is used here in the sense of *physically homogeneous*, and is applied to any system which appears of completely uniform structure throughout. The substance may be *chemically homogeneous, i.e.* it may consist entirely of the same kind of molecules. An explosive gas mixture $(2H_2 + O_2)$, or a partially dissociated vapour, both of which are chemically heterogeneous, may very well be physically homogeneous. What we specify here is that the state of the homogeneous substance is a single valued function of the temperature and the volume. Chemical transformations may take place during the processes. As long as the system is at rest, the total energy consists of the so-called *internal* energy U, which depends only on the internal state of the substance as determined by its density and temperature, and on its mass, to which it is evidently proportional. In other cases the total energy contains, besides the internal energy U, another term, namely, the kinetic energy, which is known from the principles of mechanics.

In order to determine the functional relation between U, T, and V, the state of the system must be changed, and the external effects of this change calculated. Equation (17) then gives the corresponding change of energy.

§ 68. If a gas, initially at rest and at uniform temperature, be allowed to suddenly expand by the opening of a stopcock, which makes communication with a previously exhausted vessel, a number of intricate mechanical and thermal changes will at first take place. The portion of the gas flowing into the vacuum is thrown into violent motion, then heated by impact against the sides of the vessel and by compression of the particles crowding behind, while the portion remaining in the first vessel is cooled down by expansion, etc. Assuming the walls of the vessels to be absolutely rigid and non-conducting, and denoting by 2 any particular state after communication between the vessels has been established, then, according to equation (17), the total energy of the gas in state 2 is precisely equal to that in state 1, for neither thermal nor mechanical forces have acted on the gas from without. The reaction of the walls does not perform any work. The energy in state 2 is, in general, composed of many parts, viz. the kinetic and internal energies of the gas particles, each one of which, if taken sufficiently small, may be considered as homogeneous and uniform in temperature and density. If we wait until complete rest and thermal equilibrium have been re-established, and denote this state by 2, then in 2, as in 1, the total energy consists only of the internal energy U, and we have $U_2 = U_1$. But the variables T and V, on which U depends, have passed from T_1, V_1, to T_2, V_2, where $V_2 > V_1$. By measuring the temperatures and the volumes, the relation between the temperature and the volume in processes where the internal energy remains constant may be established.

§ 69. Joule * performed such an experiment as described, and found that for perfect gases $T_2 = T_1$. He put the two communicating vessels, one filled with air at high pressure, the other exhausted, into a common water-bath at the same temperature, and found that, after the air had expanded and equilibrium had been established, the change

---

* Gay Lussac had already performed the same experiment.

of temperature of the water-bath was inappreciable. It immediately follows that, if the walls of the vessels were non-conducting, the final temperature of the total mass of the gas would be equal to the initial temperature; for otherwise the change in temperature would have communicated itself to the water-bath in the above experiment.

Hence, if the internal energy of a nearly perfect gas remains unchanged after a considerable change of volume, then its temperature also remains almost constant. In other words, *the internal energy of a perfect gas depends only on the temperature, and not on the volume.*

§ **70.** For a conclusive proof of this important deduction, much more accurate measurements are required. In Joule's experiment described above, the heat capacity of the gas is so small compared with that of the vessel and the water-bath, that a considerable change of temperature in the gas would have been necessary to produce an appreciable change of temperature in the water. More reliable results are obtained by a modification of the above method devised by Sir William Thomson (Lord Kelvin), and used by him, along with Joule, for accurate measurements. Here the outflow of the gas is artificially retarded, so that the gas passes immediately into its second state of equilibrium. The temperature T_2 is then directly measured in the stream of outflowing gas. A limited quantity of gas does not rush tumultuously into a vacuum, but a gas is slowly transferred in a steady flow from a place of high pressure, p_1, to one of low pressure, p_2 (the atmosphere), by forcing it through a boxwood tube stopped at one part of its length by a porous plug of cotton wool or filaments of silk. The results of the experiment show that when the flow has become steady there is, for air, a very small change of temperature, and, for hydrogen, a still smaller, hardly appreciable change. Hence the conclusion appears justified, that, for a perfect gas, the change of temperature vanishes entirely.

This leads to an inference with regard to the internal

energy of a perfect gas. When, after the steady state of the process has been established, a certain mass of the gas has been completely pushed through the plug, it has been operated upon by external agents during its change from the volume, V_1, at high pressure, to the larger volume, V_2, at atmospheric pressure. The mechanical equivalent of these operations, $Q + W$, is to be calculated from the external changes. No change of temperature occurs outside the tube, as the material of which it is made is practically non-conducting; hence $Q = 0$. The mechanical work done by a piston in pressing the gas through the plug at the constant pressure p_1 is evidently p_1V_1, and this for a perfect gas at constant temperature is, according to Boyle's law, equal to the work p_2V_2, which is gained by the escaping gas pushing a second piston at pressure p_2 through a volume V_2. Hence the sum of the external work W is also zero, and therefore, according to equation (17), the system, consisting of the gas and the plug, has the same total energy at the end as at the beginning, *i.e.* the difference of both total energies is zero. In this difference all terms vanish except those which depend on the volume of the gas. Also the state of the porous plug remains unaltered, and the processes which take place in it do not enter into the energy equation. There remains $U_1 - U_2 = 0$, where U_1 and U_2 denote the internal energies of the gas volumes V_1 and V_2. As the experimental results showed the temperature to be practically unchanged while the volume increased very considerably, the internal energy of a perfect gas can depend only on the temperature and not on the volume, *i.e.*,

$$\left(\frac{\partial U}{\partial V}\right)_T = 0 \quad . \quad . \quad . \quad . \quad . \quad (19)$$

For nearly perfect gases, as hydrogen, air, etc., the actual small change of temperature observed shows how far the internal energy depends on the volume. It must, however, be borne in mind that for such gases the external work,

$$W = p_1V_1 - p_2V_2,$$

does not vanish; hence the internal energy does not remain constant. For further discussion, see § 158.

§ **71.** Special theoretical importance must be attached to those thermodynamical processes which progress infinitely slowly, and which, therefore, consist of a succession of states of equilibrium. Strictly speaking, this expression is vague, since a process presupposes changes, and, therefore, disturbances of equilibrium. But where the time taken is immaterial, and the result of the process alone of consequence, these disturbances may be made as small as we please, certainly very small in comparison with the other quantities which characterize the state of the system under observation. Thus, a gas may be compressed very slowly to any fraction of its original volume, by making the external pressure, at each moment, just a trifle greater than the internal pressure of the gas. Wherever external pressure enters—as, for instance, in the calculation of the work of compression—a very small error will then be committed, if the pressure of the gas be substituted for the external pressure. On passing to the limit, even that error vanishes. In other words, the result obtained becomes rigorously exact for *infinitely slow* compression.

This holds for compression at constant as well as at variable pressure. The latter may be given the required value at each moment by the addition or removal of small weights. This may be done either by hand (by pushing the weights aside in a horizontal direction), or by means of some automatic device which acts merely as a release, and therefore does not contribute towards the work done.

§ **72.** The conduction of heat to and from the system may be treated in the same way. When it is not a question of time, but only of the amount of heat received or given out by the system, it is sufficient, according as heat is to be added to or taken from the system, to connect it with a heat-reservoir of slightly higher or lower temperature than that of the system. This small difference serves, merely, to determine the direction of the flow of the heat, while its magnitude is negligible compared with the changes of the system, which result from the process. We, therefore, speak of the conduction of heat between bodies of equal temperature, just as we speak of the compression of a gas by an external pressure equal to that of the gas. This is merely anticipating the result of passing to the limit from a small finite difference to an infinitesimal difference of temperature between the two bodies.

This applies not only to strictly isothermal processes, but also to those of varying temperature. One heat-reservoir of constant temperature will not suffice for carrying out the latter processes. These will require either an auxiliary body, the temperature of which may be arbitrarily changed, *e.g.* a gas that can be heated or cooled at pleasure by compression or expansion; or a set of constant-temperature reservoirs, each of different temperature. In the latter case, at each stage of the process we apply that particular heat-reservoir whose temperature lies nearest to that of the system at that moment.

§ **73.** The value of this method of viewing the process lies in the fact that we may imagine each *infinitely slow* process to be carried out also in the opposite direction. If a process consist of a succession of states of equilibrium with the exception of very small changes, then evidently a suitable change, quite as small, is sufficient to reverse the process. This small change will vanish when we pass over to the limiting case of the infinitely slow process, for a definite result always contains a quite definite error, and if this error be smaller than any quantity, however small, it must be zero.

§ **74.** We pass now to the application of the first law to a process of the kind indicated, and, therefore, reversible in its various parts. Taking the volume V (abscissa) and the pressure p (ordinate) as the independent variables, we may graphically illustrate our process by plotting its successive states of equilibrium in the form of a curve in the plane of the co-ordinates. Each point in this plane corresponds to a certain state of our system, the chemical nature and mass of which are supposed to be given, and each curve corresponds to a series of continuous changes of state. Let the curve α from 1 to 2 represent a reversible process which takes the substance from a state 1 to a state 2 (Fig. 2). Along α, according to equation (17), the increase of the energy is

$$U_2 - U_1 = W + Q,$$

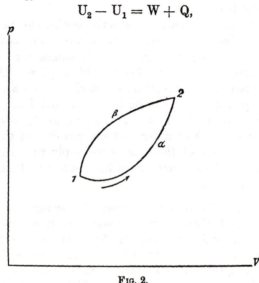

Fɪɢ. 2.

where W is the mechanical work expended on the substance, and Q the total heat absorbed by it.

§ **75.** The value of W can be readily determined. W is made up of the elementary quantities of work done on the system during the infinitesimal changes corresponding to the elements of arc of the curve α. The external pressure is at any moment equal to that of the substance, since the process is supposed to be reversible. Consequently, by the laws of hydrodynamics, the work done by the external forces in the infinitely small change is equal to the product of the pressure p, and the decrease of the volume, $- dV$, no matter what the geometrical form of the surface of the body may be. Hence the external work done during the whole process is

$$W = - \int_1^2 p dV, \quad \ldots \quad (20)$$

in which the integration extends from 1 to 2 along the curve α. If p be positive, as in the case of gases, and $V_2 > V_1$ as in Fig. 2, W is negative, *i.e.* work is not done on the system, but by the system.

In order to perform the integration, the curve α, *i.e.* the relation between p and V, must be known. As long as only the points 1 and 2 are given, the integral has no definite value. In fact, it assumes an entirely different value along a different curve, β, joining 1 and 2. Therefore $p dV$ is not a perfect differential. Mathematically this depends on the fact that p is in general not only a function of V, but also of another variable, the temperature T, which also changes along the path of integration. As long as α is not given, no statement can be made with regard to the relation between T and V, and the integration gives no definite result.

The external work, W, is evidently represented by the area (taken negative) of the plane figure bounded by the curve α, the ordinates at 1 and 2, and the axis of abscissæ. This, too, shows that W depends on the path of the curve α. Only for infinitesimal changes, *i.e.* when 1 and 2 are infinitely near one another and α shrinks to a curve element, is W determined by the initial and final points of the curve alone.

§ 76. The second measurable quantity is Q, the heat absorbed. It may be determined by calorimetric methods in calories, and then expressed in mechanical units by multiplying by the mechanical equivalent of heat. We shall now consider the theoretical determination of Q. It is, like W, the algebraical sum of the infinitely small quantities of heat added to the body during the elementary processes corresponding to the elements of the curve α. Such an increment of heat cannot, however, be immediately calculated, from the position of the curve element in the co-ordinate plane, in a manner similar to that of the increment of work. To establish an analogy between the two, one might, in imitation of the expression $- pdV$, put the increment of heat $= CdT$, where dT is the increment of temperature, and C the heat capacity, which is usually a finite quantity. But C has not, in general, a definite value. It does not depend, as the factor p in the expression for the increment of work, alone on the momentary state of the substance, *i.e.* on the position of the point of the curve considered, but also on the direction of the curve element. In isothermal changes C is evidently $= \pm \infty$, because $dT = 0$, and the heat added or withdrawn is a finite quantity. In adiabatic changes $C = 0$, for here the temperature may change in any way, while no heat is added or withdrawn. For a given point, C may, therefore, in contradistinction to p, assume all values between $+ \infty$ and $- \infty$. (*Cf.* § 47.) Hence the analogy is incomplete in one essential, and does not, in the general case, simplify the problem in hand. The splitting up of the external work into two factors p and dV is, of course, limited to the special case of reversible processes, since otherwise the external pressure is not equal to p.

§ **77.** Although the value of Q cannot, in general, be directly determined, equation (17) enables us to draw some important inferences regarding it. Substituting the value of W from equation (20) in equation (17), we obtain

$$Q = U_2 - U_1 + \int_1^2 pdV, \quad \cdot \quad \cdot \quad \cdot \quad (21)$$

which shows that the value of Q depends not only on the position of the points 1 and 2, but also, like W, on the connecting path (α or β). Carnot's theory of heat cannot be reconciled with this proposition, as we have shown at length in §§ 51 and 52.

§ **78.** The complete evaluation of Q is possible in the case where the substance returns to its initial state, having gone through a cycle of operations. This might be done by first bringing the system from 1 to 2 along α, then back from 2 to 1 along β. Then, as in all cycles (§ 65),

$$Q = -W.$$

The external work is

$$W = -\int_1^1 p\,dV,$$

the integral to be taken along the closed curve $1\alpha2\beta1$. W evidently represents the area bounded by the curve, and is positive if the process follows the direction of the arrow in Fig. 2.

§ **79.** We shall now consider the special case where the curve α, which characterizes the change of state, shrinks into an element, so that the points 1 and 2 lie infinitely near one another. W here becomes the increment of work, $-p\,dV$, and the change of the internal energy is dU. Hence, according to (21), the heat absorbed assumes the value : *

* It is usual to follow the example of Clausius, and denote this quantity by dQ to indicate that it is infinitely small. This notation, however, has frequently given rise to misunderstanding, for dQ has been repeatedly regarded as the differential of a known finite quantity Q. This faulty reasoning may be illustrated by the following example. If we choose T and V as independent variables, we have

$$d\mathrm{Q} = d\mathrm{U} + p\,d\mathrm{V} = \left(\frac{\partial \mathrm{U}}{\partial \mathrm{T}}\right)_{\mathrm{V}} d\mathrm{T} + \left[\left(\frac{\partial \mathrm{U}}{\partial \mathrm{V}}\right)_{\mathrm{T}} + p\right] d\mathrm{V}.$$

On the other hand,

$$d\mathrm{Q} = \left(\frac{\partial \mathrm{Q}}{\partial \mathrm{T}}\right)_{\mathrm{V}} d\mathrm{T} + \left(\frac{\partial \mathrm{Q}}{\partial \mathrm{V}}\right)_{\mathrm{T}} d\mathrm{V}.$$

Accordingly, since dT and dV are independent of one another,

$$\frac{\partial \mathrm{Q}}{\partial \mathrm{T}} = \frac{\partial \mathrm{U}}{\partial \mathrm{T}} \text{ and } \frac{\partial \mathrm{Q}}{\partial \mathrm{V}} = \frac{\partial \mathrm{U}}{\partial \mathrm{V}} + p.$$

Differentiating the first with respect to V and the second with respect to T, we have

$$\frac{\partial^2 \mathrm{Q}}{\partial \mathrm{T}\partial \mathrm{V}} = \frac{\partial^2 \mathrm{U}}{\partial \mathrm{T}\partial \mathrm{V}} = \frac{\partial^2 \mathrm{U}}{\partial \mathrm{T}\partial \mathrm{V}} + \frac{\partial p}{\partial \mathrm{T}},$$

therefore $\frac{\partial p}{\partial \mathrm{T}} = 0$, which is certainly not true. Such errors are excluded if we stick to the notation in the text, keeping in mind that every infinitely small quantity is not the difference of two (nearly equal) finite quantities.

$$Q = d\mathrm{U} + p d\mathrm{V}.$$

Per unit mass, this equation becomes

$$q = du + pdv, \quad . \quad . \quad . \quad . \quad (22)$$

where the small letters denote the corresponding capitals divided by M. In subsequent calculations it will often be advisable to use T as an independent variable, either in conjunction with p, or with v. We shall, in each case, select as independent variables those which are most conducive to a simplification of the problem in hand. The meaning of the differentiation will be indicated whenever a misunderstanding is possible.

We shall now apply our last equation (22) to the most important reversible processes.

§ 80. It has been repeatedly mentioned that the specific heat of a substance may be defined in very different ways according to the manner in which the heating is carried out. But, according to § 46 and equation (22), we have, for any heating process,

$$c = \frac{q}{d\mathrm{T}} = \frac{du}{d\mathrm{T}} + p\frac{dv}{d\mathrm{T}}. \quad . \quad . \quad (23)$$

In order to give a definite meaning to the differential coefficients, some arbitrary condition is required, which will prescribe the direction of the change. A single condition is sufficient, since the state of the substance depends on two variables only.

§ 81. Heating at Constant Volume.—Here $dv = 0$, $c = c_v$, the specific heat at constant volume. Hence, according to equation (23),

$$c_v = \left(\frac{\partial u}{\partial \mathrm{T}}\right)_v \quad . \quad . \quad . \quad . \quad (24)$$

or

$$c_v = \left(\frac{\partial u}{\partial p}\right)_v\left(\frac{\partial p}{\partial \mathrm{T}}\right)_v \quad . \quad . \quad . \quad (25)$$

§ 82. Heating under Constant Pressure.—Here

$dp = 0$, $c = c_p$, the specific heat at constant pressure. According to equation (23),

$$c_p = \left(\frac{\partial u}{\partial T}\right)_p + p\left(\frac{\partial v}{\partial T}\right)_p \quad \cdots \quad (26)$$

or

$$c_p = \left[\left(\frac{\partial u}{\partial v}\right)_p + p\right]\left(\frac{\partial v}{\partial T}\right)_p \quad \cdots \quad (27)$$

By the substitution of

$$\left(\frac{\partial u}{\partial T}\right)_p = \left(\frac{\partial u}{\partial T}\right)_v + \left(\frac{\partial u}{\partial v}\right)_T\left(\frac{\partial v}{\partial T}\right)_p$$

in (26), c_p may be written in the form

$$c_p = \left(\frac{\partial u}{\partial T}\right)_v + \left[\left(\frac{\partial u}{\partial v}\right)_T + p\right]\left(\frac{\partial v}{\partial T}\right)_p,$$

or, by (24),

$$c_p = c_v + \left[\left(\frac{\partial u}{\partial v}\right)_T + p\right]\left(\frac{\partial v}{\partial T}\right)_p \quad \cdots \quad (28)$$

§ 83. By comparing (25) and (27) and eliminating u, we are led to a direct experimental test of the theory.

By (25), $\qquad \left(\frac{\partial u}{\partial p}\right)_v = c_v\left(\frac{\partial T}{\partial p}\right)_v,$

and by (27), $\qquad \left(\frac{\partial u}{\partial v}\right)_p = c_p\left(\frac{\partial T}{\partial v}\right)_p - p\,;$

whence, differentiating the former equation with respect to v, keeping p constant, and the latter with respect to p, keeping v constant, and equating, we have

$$\frac{\partial}{\partial v}\left(c_v\frac{\partial T}{\partial p}\right) = \frac{\partial}{\partial p}\left(c_p\frac{\partial T}{\partial v} - p\right)$$

or $\qquad (c_p - c_v)\frac{\partial^2 T}{\partial p\partial v} + \frac{\partial c_p}{\partial p}\cdot\frac{\partial T}{\partial v} - \frac{\partial c_v}{\partial v}\cdot\frac{\partial T}{\partial p} = 1 \quad \cdots \quad (29)$

This equation contains only quantities which may be experimentally determined, and therefore furnishes a means for testing the first law of thermodynamics by observations on any homogeneous substance.

§ 84. Perfect Gases.—The above equations undergo considerable simplifications for perfect gases. We have, from (14),

$$p = \frac{R}{m} \cdot \frac{T}{v}, \quad \ldots \quad \ldots \quad (30)$$

where $R = 8\cdot315 \times 10^7$ and m is the (real or apparent) molecular weight. Hence

$$T = \frac{m}{R} pv,$$

and equation (29) becomes

$$c_p - c_v + p\frac{\partial c_p}{\partial p} - v\frac{\partial c_v}{\partial v} = \frac{R}{m}.$$

Assuming that only the laws of Boyle, Gay Lussac, and Avogadro hold, no further conclusions can be drawn from the first law of thermodynamics with regard to perfect gases.

§ 85. We shall now make use of the additional property of perfect gases, established by Thomson and Joule (§ 70), that the internal energy of a perfect gas depends only on the temperature, and not on the volume, and that hence per unit mass, according to (19),

$$\left(\frac{\partial u}{\partial v}\right)_T = 0 \quad . \quad \ldots \quad \ldots \quad (31)$$

The general equation,

$$du = \left(\frac{\partial u}{\partial T}\right)_v dT + \left(\frac{\partial u}{\partial v}\right)_T dv,$$

then becomes, for perfect gases,

$$du = \left(\frac{\partial u}{\partial T}\right)_v dT,$$

and, according to (24),

$$du = c_v \cdot dT \quad \ldots \quad \ldots \quad (32)$$

It follows from (28) that

$$c_p = c_v + p\left(\frac{\partial v}{\partial T}\right)_p,$$

or, considering the relation (30),

$$c_p = c_v + \frac{R}{m};$$

i.e. the difference between the specific heat at constant pressure and the specific heat at constant volume is constant. We have, therefore,

$$mc_p - mc_v = R \quad . \quad . \quad . \quad . \quad (33)$$

The difference between the molecular heats at constant pressure and constant volume is, therefore, independent even of the nature of the gas.

§ **86.** Only the specific heat at constant pressure, c_p, lends itself to simple experimental determination,* because a quantity of gas enclosed in a vessel of constant volume has too small a heat capacity to produce considerable thermal effects on the surrounding bodies. Since c_v, according to (24), like u, depends on the temperature only, and not on the volume, the same follows for c_p, according to (33). This conclusion was first confirmed by Regnault's experiments. He found c_p constant within a considerable range of temperature. By (33), c_v is constant within the same range. We shall, therefore, extend the definition of ideal gases so that c_p and c_v are completely independent of temperature and volume.

If the molecular heats be expressed in calories, R must be divided by Joule's mechanical equivalent of heat J. The difference between the molecular heats at constant pressure and at constant volume is then

$$mc_p - mc_v = \frac{R}{J} = \frac{8 \cdot 315 \times 10^7}{4 \cdot 19 \times 10^7} = 1 \cdot 985 \quad . \quad . \quad (34)$$

§ **87.** The following table contains the specific heats and molecular heats of several gases at constant pressure, measured by direct experiment; also the molecular heats

---

* By an improved form of steam calorimeter Joly was able to make direct experimental determinations of the specific heats of gases at constant volume.—TR.

at constant volume found by subtracting 1·985 therefrom, and also the ratio $\frac{c_p}{c_v} = \gamma$:

	c_p Specific heat at const. pressure.	m Molecular weight.	mc_p Molecular heat at const. pressure.	mc_v Molecular heat at const. volume.	$\frac{c_p}{c_v} = \gamma$.
Hydrogen . .	3·410	2·016	6·87	4·88	1·41
Oxygen . .	0·220	32·0	7·04	5·05	1·40
Nitrogen . .	0·2438	28·0	6·83	4·85	1·41
Air	0·2404	28·9	6·95	4·96	1·40

The specific heats of all these gases increase slowly if the temperature be considerably increased. Within the range of temperature in which the specific heat is constant, equation (32) can be integrated, giving

$$u = c_v \mathrm{T} + \text{const.} \quad . \quad . \quad . \quad . \quad (35)$$

The constant of integration depends on the selection of the zero point of energy. For perfect gases, as we have laid down in § 86, c_p and c_v are constant throughout, hence the last equation holds good in general.

§ 88. Adiabatic Process.—The characteristic feature of the adiabatic process is that $q = 0$, and, according to equation (22),

$$0 = du + p\,dv.$$

Assuming, again, a perfect gas, and substituting the values of du from (32) and of p from (30), we have

$$0 = c_v d\mathrm{T} + \frac{\mathrm{R}}{m} \cdot \frac{\mathrm{T}}{v} dv, \quad . \quad . \quad . \quad (36)$$

or, on integrating,

$$c_v \log \mathrm{T} + \frac{\mathrm{R}}{m} \log v = \text{const.}$$

Replacing $\frac{\mathrm{R}}{m}$ according to (33) by $c_p - c_v$, and dividing by c_v, we get

$$\log \mathrm{T} + (\gamma - 1) \log v = \text{const.} \quad . \quad . \quad . \quad (37)$$

(*i.e.* during adiabatic expansion the temperature falls).

Remembering that according to the characteristic equation (30)

$$\log p + \log v - \log T = \text{const.,}$$

we have, on eliminating v,

$$- \gamma \log T + (\gamma - 1) \log p = \text{const.}$$

(*i.e.* during adiabatic compression the temperature rises), or, on eliminating T,

$$\log p + \gamma \log v = \text{const.}$$

The values of the constants of integration are given by the initial state of the process.

If we compare our last equation in the form

$$pv^\gamma = \text{const.} \quad \ldots \quad \ldots \quad (38)$$

with Boyle's law $pv = \text{const.}$, it is seen that during adiabatic compression the volume decreases more slowly for an increase of pressure than during isothermal compression, because during adiabatic compression the temperature rises. The adiabatic curves in the pv — plane (§ 22) are, therefore, steeper than the hyperbolic isotherms.

§ 89. Adiabatic processes may be used in various ways for the determination of γ, the ratio of the specific heats. The agreement of the results with the value calculated from the mechanical equivalent of heat forms an important confirmation of the theory.

Thus, the measurement of the velocity of sound in a gas may be used for determining the value of γ. It is proved in hydrodynamics that the velocity of sound in a fluid is $\sqrt{\dfrac{dp}{d\rho}}$, where $\rho = \dfrac{1}{v}$, the density of the fluid. Since gases are bad conductors of heat, the compressions and expansions which accompany sound-vibrations must be considered as adiabatic, and not isothermal, processes. The relation between the pressure and the density is, therefore, in the case of perfect gases, not that expressed by Boyle's law $\dfrac{p}{\rho} = pv = \text{const.}$, but that given by equation (38), viz.:

$$\frac{p}{\rho^\gamma} = \text{const.}$$

Hence, by differentiation,

$$\frac{dp}{d\rho} = \frac{\gamma p}{\rho} = \gamma p v,$$

or, according to (30),

$$\frac{dp}{d\rho} = \gamma \frac{R}{m} T,$$

$$\gamma = \frac{m}{RT} \cdot \frac{dp}{d\rho}.$$

In air at $0°$, the velocity of sound is $\sqrt{\dfrac{dp}{d\rho}} = 33170 \dfrac{cm.}{sec.}$; hence, according to our last equation, taking the values of m from § 41, and of R from § 84, and $T = 273$,

$$\gamma = \frac{28 \cdot 9}{8 \cdot 315 \times 10^7} \cdot \frac{33170^2}{273} = 1 \cdot 40.$$

This agrees with the value calculated in § 87.

Conversely, the value of γ, calculated from the velocity of sound, may be used in the calculation of c_v in calories, for the determination of the mechanical equivalent of heat from (33). This method of evaluating the mechanical equivalent of heat was first proposed by Robert Meyer in 1842. It is true that the assumption expressed in equation (31), that the internal energy of air depends only on the temperature, is essential to this method, or in other words, that the difference of the specific heats at constant pressure and constant volume depends only on the external work. The direct proof of this fact, however, must be considered as being first given by the experiments of Thomson and Joule, described in § 70.

§ 90. We shall now consider a more complex process, a reversible cycle of a special kind, which has played an important part in the development of thermodynamics, known as Carnot's cycle, and shall apply the first law to it in detail.

Let a substance of unit mass, starting from an initial state characterized by the values T_1, v_1, first be compressed *adiabatically* until its temperature rises to $T_2(T_2 > T_1)$ and its volume is reduced to $v_2(v_2 < v_1)$ (Fig. 3). Second, suppose it be now allowed to expand *isothermally* to volume $v_2'(v_2' > v_2)$, in constant connection with a heat-reservoir of constant temperature, T_2, which gives out the heat of expansion Q_2. Third, let it be further expanded *adiabatically* until its temperature falls to T_1, and the volume is thereby increased to

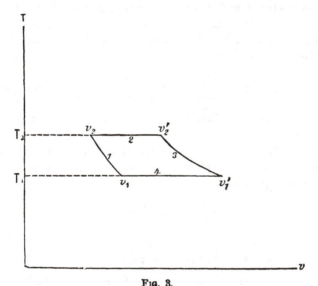

Fɪɢ. 3.

v_1'. Fourth, let it be compressed *isothermally* to the original volume v_1, while a heat-reservoir maintains the temperature at T_1, by absorbing the heat of compression. All these operations are to be carried out in the reversible manner described in § 71. The sum of the heat absorbed by the system, and the work done on the system during this cycle is, by the first law,

$$Q + W = 0 \quad . \quad . \quad . \quad . \quad (39)$$

The heat Q, that has been absorbed by the substance, is

$$Q = Q_1 + Q_2 \quad . \quad . \quad . \quad . \quad (40)$$

(Q_1 is here negative). The external work W may be calculated from the adiabatic and the isothermal compressibility of the substance. According to (20),

$$W = - \int_{v_1, T_1}^{v_2, T_2} p\,dv - \int_{v_2, T_2}^{v_2', T_2} p\,dv - \int_{v_2', T_2}^{v_1', T_1} p\,dv - \int_{v_1', T_1}^{v_1, T_1} p\,dv.$$

These integrals are to be taken along the curves 1, 2, 3, 4 respectively; 1 and 3 being adiabatic, 2 and 4 isothermal.

Assuming the substance to be a perfect gas, the above integrals can readily be found. If we bear in mind the relations (30) and (36), we have

$$W = \int_{T_1}^{T_2} c_v\,dT - \frac{R}{m}\int_{v_2}^{v_2'} \frac{T_2}{v}\,dv + \int_{T_2}^{T_1} c_v\,dT - \frac{R}{m}\int_{v_1'}^{v_1} \frac{T_1}{v}\,dv \quad (41)$$

The work of the adiabatic compression in the first part of the process is equal in value and opposite in sign to that of the adiabatic expansion in the third part of the process. There remains, therefore, the sum of the work in the isothermal portions:

$$W = - \frac{R}{m}\Big(T_2 \log \frac{v_2'}{v_2} + T_1 \log \frac{v_1}{v_1'}\Big).$$

Now, the state (v_2, T_2) was developed from (v_1, T_1) by an adiabatic process; therefore, by (37),

$$\log T_2 + (\gamma - 1) \log v_2 = \log T_1 + (\gamma - 1) \log v_1.$$

Similarly, for the adiabatic process, which leads from (v_2', T_2) to (v_1', T_1),

$$\log T_2 + (\gamma - 1) \log v_2' = \log T_1 + (\gamma - 1) \log v_1'.$$

From these equations, it follows that

$$\frac{v_2'}{v_2} = \frac{v_1'}{v_1},$$

and $$\therefore W = - \frac{R}{m}(T_2 - T_1) \log \frac{v_1'}{v_1}.$$

Since, in the case considered, $T_2 > T_1$, and $\dfrac{v_1'}{v_1} = \dfrac{v_2'}{v_2} > 1$, the total external work W is negative, *i.e.* mechanical work has been gained by the process. But, from (39) and (40),

$$Q = Q_1 + Q_2 = -W; \quad . \quad . \quad . \quad (42)$$

therefore Q is positive, *i.e.* the heat-reservoir at temperature T_2 has lost more heat than the heat-reservoir at temperature T_1 has gained.

The value of W, substituted in the last equation, gives

$$Q = Q_1 + Q_2 = \frac{R}{m}(T_2 - T_1) \log \frac{v_1'}{v_1} . \quad . \quad (43)$$

The correctness of this equation is evident from the direct calculation of the values of Q_1 and Q_2. The gas expands isothermally while the heat-reservoir at temperature T_2 is in action. The internal energy of the gas therefore remains constant, and the heat absorbed is equal in magnitude and opposite in sign to the external work. Hence, by equating Q_2 to the second integral in (41),

$$Q_2 = \frac{R}{m}T_2 \log \frac{v_2'}{v_2} = \frac{R}{m}T_2 \log \frac{v_1'}{v_1},$$

and, similarly, by equating Q_1 to the fourth integral in (41),

$$Q_1 = \frac{R}{m}T_1 \log \frac{v_1}{v_1'} = -\frac{R}{m}T_1 \log \frac{v_1'}{v_1},$$

which agrees with equation (43).

There exists, then, between the quantities Q_1, Q_2, W, besides the relation given in (42), this new relation :

$$Q_1 : Q_2 : W = (-T_1) : T_2 : (T_1 - T_2) \quad . \quad (44)$$

§ **91.** The Carnot cycle just described, if sufficiently often repeated, forms a type of cyclic machine by which heat may be continuously converted into work. In order to survey this in detail, we shall now examine more closely the result of all the interreactions which take place on carrying out such a cyclic process. To this end we shall compare the initial and final states of all the bodies concerned. The gas operated upon has not been changed in any way by the process, and may be left out of account. It has done service only as a transmitting agent, in order to bring about changes in the surroundings. The two reservoirs, however, have undergone a change, and, besides, a positive amount of external work, $W' = -W$, has been gained; *i.e.* at the close of the process certain weights, which were in action during the compression and the expansion, are found to be at a higher level than at the beginning, or a spring, serving a similar purpose, is at a greater tension. On the other hand, the heat-reservoir at T_2 has given out heat to the amount Q_2, and the cooler reservoir at T_1 has received the smaller amount $Q_1' = -Q_1$. The heat that has vanished is equivalent to the work gained. This result may be briefly expressed as follows: The quantity of heat Q_2, at temperature T_2, has passed in part (Q_1') to a lower temperature (T_1), and has in part $(Q_2 - Q_1' = Q_1 + Q_2)$ been transformed into mechanical work. Carnot's cycle, performed with a perfect gas, thus affords a means of drawing heat from a body and of gaining work in its stead, without introducing any changes in nature except the transference of a certain quantity of heat from a body of higher temperature to one of lower temperature.

But, since the process described is reversible in all its parts, it may be put into effect in such a way that all the quantities, Q_1, Q_2, W, change sign, Q_1 and W becoming positive, $Q_2 = -Q_2'$ negative. In this case the hotter reservoir at T_2 receives heat to the amount Q_2', partly from the colder reservoir (at T_1), and partly from the mechanical work expended (W). By reversing Carnot's cycle, we have, then, a means of transferring heat from a colder to a hotter body without introducing any other changes in nature than the transformation of a certain amount of mechanical work into heat. We shall see, later, that, for the success of Carnot's reversible cycle, the nature of the transmitting agent or working substance is immaterial, and that perfect gases are, in this respect, neither superior nor inferior to other substances (*cf.* § 137).

CHAPTER III.

APPLICATIONS TO NON-HOMOGENEOUS SYSTEMS.

§ 92. THE propositions discussed in the preceding chapter are, in a large part, also applicable to substances which are not perfectly homogeneous in structure. We shall, therefore, in this chapter consider mainly such phenomena as characterize the inhomogeneity of a system.

Let us consider a system composed of a number of homogeneous bodies in juxtaposition, separated by given bounding surfaces. Such a system may, or may not, be chemically homogeneous. A liquid in contact with its vapour is an example of the first case, if the molecules of the latter be identical with those of the former. A substance in contact with another of different chemical constitution, is an example of the second. Whether a system is physically homogeneous or not, can, in most cases, be ascertained beyond doubt, by finding surfaces of contact within the system, optically or by other means—in the case of emulsions, for example, by determining the vapour pressure or the freezing point. The question as to the chemical homogeneity, *i.e.* the presence of one kind of molecule only, is much more difficult, and has hitherto been answered only in special cases. For this reason we classify substances according to their physical and not according to their chemical homogeneity.

§ 93. One characteristic of processes in non-homogeneous systems consists in their being generally accompanied by considerable changes of temperature, *e.g.* in evaporation or in oxidation. To maintain the initial temperature and pressure consequently requires considerable exchange of heat with the surroundings and corresponding external work. The latter, however, is generally small compared with the external heat, and may be neglected in most chemical processes. In thermochemistry, therefore, the external effects,

$$Q + W = U_2 - U_1, \quad \ldots \quad (45)$$

are generally measured in calories (the heat equivalent of the external effects). The external work, W, is small compared with Q. Furthermore, most chemical processes are accompanied by a rise in temperature, or, if the initial temperature be re-established, by an external yield of heat (exothermal processes). Therefore, in thermochemistry, the heat given out to the surroundings in order to restore the initial temperature is denoted as the " positive heat of reaction " of the process. In our equations we shall therefore use Q (the heat absorbed) with the negative sign, in processes with positive heat of reaction (*e.g.* combustion); with the positive sign, in those with negative heat of reaction (*e.g.* evaporation, fusion, dissociation).

§ **94.** To make equation (45) suitable for thermochemistry it is expedient to denote the internal energy U of a system in a given state, by a symbol denoting its chemical constitution. J. Thomsen introduced a symbol of this kind. He denoted by the formulæ for the atomic or molecular weight of the substances enclosed in brackets, the internal energy of a corresponding weight referred to an arbitrary zero of energy. Through W. Ostwald this notation has come into general use. Thus [Pb], [S], [PbS] denote the energies of an atom of lead, an atom of sulphur, and a molecule of lead sulphide respectively. In order to express the fact that the formation of a molecule of lead sulphide from its atoms is accompanied by an evolution of heat of 18,400 cal., the external work of the process being negligible, we put

$$U_1 = [Pb] + [S]; \ U_2 = [PbS];$$
$$W = 0; \ Q = -18,400 \text{ cal.,}$$

and equation (45) becomes

$$- 18{,}400 \text{ cal.} = [\text{PbS}] - [\text{Pb}] - [\text{S}],$$

or, as usually written,

$$[\text{Pb}] + [\text{S}] - [\text{PbS}] = 18{,}400 \text{ cal.}$$

This means that the internal energy of lead and sulphur, when separate, is 18,400 calories greater than that of their combination at the same temperature. The use of the molecular formulæ is a check that the energies, which are being compared, refer to the same material system. The equation could be simplified by selecting the uncombined state of the elements Pb and S as the zero of energy. Then (§ 64), $[\text{Pb}] = 0$ and $[\text{S}] = 0$, and

$$[\text{PbS}] = - 18{,}400 \text{ cal.}$$

§ 95. To define accurately the state of a substance, and thereby its energy, besides its chemical nature and mass, its temperature and pressure must be given. If no special statement is made, as in the above example, mean laboratory temperature, *i.e.* about 18° C., is generally assumed, and the pressure is supposed to be atmospheric pressure. The pressure has, however, very little influence on the internal energy; in fact, none at all in the case of perfect gases [equation (35)].

The state of aggregation should also be indicated. This may be done, where necessary, by using brackets for the solids, parentheses for liquids, and braces for gases. Thus $[\text{H}_2\text{O}]$, (H_2O), $\{\text{H}_2\text{O}\}$ denote the energies of a molecule of ice, water, and water vapour respectively. Hence, for the fusion of ice at 0° C.,

$$(\text{H}_2\text{O}) - [\text{H}_2\text{O}] = 80 \times 18 = 1440 \text{ cal.}$$

It is often desirable, as in the case of solid carbon, sulphur, arsenic, or isomeric compounds, to denote by some means the special modification of the substance.

These symbols may be treated like algebraic quantities, whereby considerations, which would otherwise present considerable complications, may be materially shortened. Examples of this are given below.

§ **96.** To denote the energy of a solution or mixture of several compounds, we may write the formulæ for the molecular weights with the requisite number of molecules. Thus,

$$(H_2SO_4) + 5(H_2O) - (H_2SO_4 . 5H_2O) = 13,100 \text{ cal.}$$

means that the solution of 1 molecule of sulphuric acid in 5 molecules of water gives out 13,100 calories of heat. Similarly, the equation

$$(H_2SO_4) + 10(H_2O) - (H_2SO_4 . 10H_2O) = 15,100 \text{ cal.}$$

gives the heat of the solution of the same in 10 molecules of water. By subtracting the first equation from the second, we get

$$(H_2SO_4 . 5H_2O) + 5(H_2O) - (H_2SO_4 . 10H_2O) = 2000 \text{ cal.,}$$

i.e. on diluting a solution of 1 molecule of sulphuric acid dissolved in 5 molecules of water, by the addition of another 5 molecules of water, 2000 calories are given out.

§ **97.** As a matter of experience, in very dilute solutions further dilution no longer yields any appreciable amount of heat. Thus, in indicating the internal energy of a dilute solution it is often unnecessary to give the number of molecules of the solvent. We write briefly

$$(H_2SO_4) + (aq.) - (H_2SO_4 \text{ aq.}) = 17,900 \text{ cal.}$$

to express the heat of reaction of infinite dilution of a molecule of sulphuric acid. Here (aq.) denotes any amount of water sufficient for the practical production of an infinitely dilute solution.

§ **98.** Volumetric changes being very slight in chemical processes which involve only solids and liquids, the heat equivalent of the external work W (§ 93) is a negligible quantity compared with the heat of reaction. The latter alone, then, represents the change of energy of the system :

$$U_2 - U_1 = Q.$$

It, therefore, depends on the initial and final states only, and not on the intermediate steps of the process. These considerations do not apply, in general, when gaseous substances enter into the reaction. It is only in the combustions in the " calorimetric bomb," extensively used by Berthelot and Stohmann in their investigations, that the volume remains constant and the external work is zero. In these reactions the heat of reaction observed represents the total change of energy. In other cases, however, the amount of external work W may assume a considerable value, and it is materially influenced by the process itself. Thus, a gas may be allowed to expand, at the same time performing work, which may have any value within certain limits, from zero upwards. But since its change of energy $U_2 - U_1$ depends on the initial and final states only, a greater amount of work done against the external forces necessitates a smaller heat of reaction for the process, and *vice versâ*. To find the latter, not only the change of the internal energy, but also the amount of the external work must be known. This renders necessary an account of the external conditions under which the process takes place.

§ **99.** Of all the external conditions that may accompany a chemical process, constant (atmospheric) pressure is the one which is of the most practical importance : $p = p_0$. The external work is then, according to equation (20),

$$W = - \int_1^2 p_0 dV = p_0(V_1 - V_2); \quad . \quad . \quad (46)$$

that is, equal to the product of the pressure and the decrease of volume. This, according to (45), gives

$$U_2 - U_1 = Q + p_0(V_1 - V_2). \quad . \quad . \quad (47)$$

Now, the total decrease of volume, $V_1 - V_2$, may generally be put equal to the decrease of volume of the gaseous portions of the system, neglecting that of the solids and liquids. Since, by (16),

$$V_1 - V_2 = R\frac{T}{p_0}(n_1 - n_2),$$

where n_1, n_2 are the number of gas molecules present before and after the reaction, the heat equivalent of the external work at constant pressure is, by (46) and (34),

$$\frac{W}{J} = \frac{p_0(V_1 - V_2)}{J} = \frac{R}{J}T(n_1 - n_2) = 1{\cdot}985T(n_1 - n_2)\ \text{cal.}$$

The heat of reaction of a process at constant pressure is therefore

$$-Q = U_1 - U_2 + 1{\cdot}985T(n_1 - n_2)\ \text{cal.} \quad . \quad (48)$$

If, for instance, one gram molecule of hydrogen and half a gram molecule of oxygen, both at 18° C., combine at constant pressure to form water at 18° C., we put

$$U_1 = \{H_2\} + \tfrac{1}{2}\{O_2\};\ U_2 = (H_2O);\ n_1 = \tfrac{3}{2};\ n_2 = 0;\ T = 291.$$

The heat of combustion is, therefore, by (48),

$$-Q = \{H_2\} + \tfrac{1}{2}\{O_2\} - (H_2O) + 860\ \text{cal.},$$

i.e. 860 cal. more than would correspond to the decrease of the internal energy, or to the combustion without the simultaneous performance of external work.

§ **100.** If we write equation (47) in the form

$$(U + p_0V)_2 - (U + p_0V)_1 = Q, \quad . \quad . \quad (49)$$

it will be seen that, in processes under constant pressure p_0, the heat of reaction depends only on the initial and final states, just as in the case when there is no external work. The heat of reaction, however, is not equal to the difference of the internal energies U, but to the difference of the values of the quantity

$$U + pV = H$$

at the beginning and end of the process. Gibbs has called

the quantity H the *heat function at constant pressure*. This function plays exactly the same part with respect to the heat of reaction of isobaric processes as the energy does with respect to the heat of reaction of isochoric processes.

If, then, only processes at constant pressure be considered, it will be expedient to regard the symbols $\{H_2\}$, $\{H_2O\}$, etc., as representing the heat function H instead of simply the energy U. Thus the difference in the two values of the function will, in all cases, directly represent the heat of the reaction. This notation is therefore adopted in the following.

§ **101.** To determine the heat of the reaction of a chemical reaction at constant pressure, the initial and final values of the heat function, H, of the system suffice. The general solution of this problem, therefore, amounts to finding the heat functions of all imaginable material systems in all possible states. Frequently, different ways of transition from one state of a system to another may be devised, which may serve either as a test of the theory, or as a check upon the accuracy of the observations. Thus J. Thomsen found the heat of neutralization of a solution of sodium bicarbonate with caustic soda to be :

$$(NaHCO_3 \text{ aq.}) + (NaHO \text{ aq.}) - (Na_2CO_3 \text{ aq.}) = 9200 \text{ cal.}$$

He also found the heat of neutralization of carbon dioxide to be :

$$(CO_2 \text{ aq.}) + 2(NaHO \text{ aq.}) - (Na_2CO_3 \text{ aq.}) = 20.200 \text{ cal.}$$

By subtraction

$$(CO_2 \text{ aq.}) + (NaHO \text{ aq.}) - (NaHCO_3 \text{ aq.}) = 11,000 \text{ cal.}$$

This is the heat of reaction corresponding to the direct combination of carbon dioxide and caustic soda to form sodium bicarbonate. Berthelot verified this by direct measurement.

§ **102.** Frequently, of two ways of transition, one is better adapted for calorimetric measurements than the other. Thus, the heat developed by the decomposition of hydrogen peroxide into water and oxygen cannot readily be measured directly. Thomsen therefore oxidized a solution of stannous chloride in hydrochloric acid first by means of hydrogen peroxide :

$$(SnCl_2 \quad 2HCl \text{ aq.}) + (H_2O_2 \text{ aq.}) - (SnCl_4 \text{ aq.}) = 88{,}800 \text{ cal.},$$

then by means of oxygen gas :

$$(SnCl_2 . 2HCl \text{ aq.}) + \tfrac{1}{2}\{O_2\} - (SnCl_4 \text{ aq.}) = 65{,}700 \text{ cal.}$$

Subtraction gives

$$(H_2O_2 \text{ aq.}) - \tfrac{1}{2}\{O_2\} - (\text{aq.}) = 23{,}100 \text{ cal.}$$

for the heat of reaction on the decomposition of dissolved hydrogen peroxide into oxygen and water.

§ **103.** The *heat of formation* of carbon monoxide from solid carbon and oxygen cannot be directly determined. because carbon never burns completely to carbon monoxide, but always, in part, to carbon dioxide as well. Therefore Favre and Silbermann determined the heat of reaction on the complete combustion of carbon to carbon dioxide :

$$[C] + \{O_2\} - \{CO_2\} = 97{,}000 \text{ cal.},$$

and then determined the heat of reaction on the combustion of carbon monoxide to carbon dioxide :

$$\{CO\} + \tfrac{1}{2}\{O_2\} - \{CO_2\} = 68{,}000 \text{ cal.}$$

By subtraction we get

$$[C] + \tfrac{1}{2}\{O_2\} - \{CO\} = 29{,}000 \text{ cal.},$$

the required heat of formation of carbon monoxide.

§ **104.** According to the above, theory enables us to calculate the heat of reaction of processes which cannot be directly realized, for as soon as the heat function of a system has been found in any way, it may be compared with other heat functions.

Let the problem be, *e.g.*, to find the heat of formation of liquid carbon bisulphide from solid carbon and solid sulphur, which do not combine directly. The following represent the reactions :—

The combustion of solid sulphur to sulphur dioxide gas :

$$[S] + \{O_2\} - \{SO_2\} = 71,100 \text{ cal.}$$

The combustion of solid carbon to carbon dioxide :

$$[C] + \{O_2\} - \{CO_2\} = 97,000 \text{ cal.}$$

The combustion of carbon bisulphide vapour to carbon dioxide and sulphur dioxide :

$$\{CS_2\} + 3\{O_2\} - \{CO_2\} - 2\{SO_2\} = 265,100 \text{ cal.}$$

The condensation of carbon bisulphide vapour :

$$\{CS_2\} - (CS_2) = 6400 \text{ cal.}$$

Elimination by purely arithmetical processes furnishes the required heat of formation :

$$[C] + 2[S] - (CS_2) = -19,500 \text{ cal.,}$$

hence negative.

In organic thermochemistry the most important method of determining the heat of formation of a compound consists in determining the heat of combustion, first of the compound, and then of its constituents.

Methane (marsh gas) gives by the complete combustion to carbon dioxide and water (liquid) :

$$\{CH_4\} + 2\{O_2\} - \{CO_2\} - 2(H_2O) = 211,900 \text{ cal.,}$$
but
$$\{H_2\} + \tfrac{1}{2}\{O_2\} - (H_2O) = 68,400 \text{ cal.,} \qquad (50)$$
and
$$[C] + \{O_2\} - \{CO_2\} = 97,000 \text{ cal.;}$$

therefore, by elimination, we obtain the heat of formation of methane from solid carbon and hydrogen gas :

$$[C] + 2\{H_2\} - \{CH_4\} = 21,900 \text{ cal.}$$

§ **105.** The external heat, Q, of a given change at constant pressure will depend on the temperature at which the process is carried out. In this respect the first law of thermodynamics leads to the following relation :

From equation (49) it follows that, for any temperature T,

$$H_2 - H_1 = Q,$$

for another temperature T',

$$H_2' - H_1' = Q',$$

and by subtraction

$$Q' - Q = (H_2' - H_2) - (H_1' - H_1),$$

i.e. the difference of the heats of reaction (Q — Q') is equal to the difference in the quantities of heat, which before and after the reaction would be required to raise the temperature of the system from T to T'.

If we take T' — T infinitely small, we have

$$\frac{dQ}{dT} = \left(\frac{dH}{dT}\right)_2 - \left(\frac{dH}{dT}\right)_1 = C_2 - C_1 \quad . \quad . \quad (50a)$$

where C_1 and C_2 are the heat capacities before and after the transformation.

Thus the influence of the temperature on the combustion of hydrogen to water (liquid) may be found by comparing the heat capacity of the mixture $(H_2 + \frac{1}{2}O_2)$ with that of the water (H_2O). The former is equal to the molecular heat of hydrogen plus half the molecular heat of oxygen. According to the table in § 87, this is

$$C_1 = 6.87 + 3.52 = 10.39$$

and

$$C_2 = 1 \times 18 = 18.$$

Hence, by (50a),

$$\frac{dQ}{dT} = 7.6.$$

Since Q is negative, the heat of combustion of a gram molecule of hydrogen decreases with rising temperature by 7·6 cal. per degree Centigrade.

PART III.

The Second Fundamental Principle of Thermodynamics.

CHAPTER I.

INTRODUCTION.

§ **106.** The second law of thermodynamics is essentially different from the first law, since it deals with a question in no way touched upon by the first law, viz. the direction in which a process takes place in nature. Not every change which is consistent with the principle of the conservation of energy satisfies also the additional conditions which the second law imposes upon the processes, which actually take place in nature. In other words, the principle of the conservation of energy does not suffice for a unique determination of natural processes.

If, for instance, an exchange of heat by conduction takes place between two bodies of different temperature, the first law, or the principle of the conservation of energy, merely demands that the quantity of heat given out by the one body shall be equal to that taken up by the other. Whether the flow of heat, however, takes place from the colder to the hotter body, or *vice versâ*, cannot be answered by the energy principle alone. The very notion of temperature is alien to that principle, as can be seen from the fact that it yields no exact definition of temperature. Neither does the general equation (17) of the first law contain any statement with regard to the direction of the particular process. The special equation (50), for instance,

$$\{H_2\} + \tfrac{1}{2}\{O_2\} - (H_2O) = 68{,}400 \text{ cal.,}$$

means only that, if hydrogen and oxygen combine under constant pressure to form water, the re-establishment of the initial temperature requires a certain amount of heat to be given up to surrounding bodies; and *vice versâ*, that this amount of heat is absorbed when water is decomposed into hydrogen and oxygen. It offers no information, however, as to whether hydrogen and oxygen actually combine to form water, or water decomposes into hydrogen and oxygen, or whether such a process can take place at all in either direction. From the point of view of the first law, the initial and final states of any process are completely equivalent.

§ **107.** In one particular case, however, does the principle of the conservation of energy prescribe a certain direction to a process. This occurs when, in a system, one of the various forms of energy is at an absolute maximum, or absolute minimum. It is evident that, in this case, the direction of the change must be such that the particular form of energy will decrease, or increase. This particular case is realized in mechanics by a system of particles at rest. Here the kinetic energy is at an absolute minimum, and, therefore, any change of the system is accompanied by an increase of the kinetic energy, and, if it be an isolated system, by a decrease of the potential energy. This gives rise to an important proposition in mechanics, which characterizes the direction of possible motion, and lays down, in consequence, the general condition of mechanical equilibrium. It is evident that, if both the kinetic and potential energies be at a minimum, no change can possibly take place, since none of these can increase at the expense of the other. The system must, therefore, remain at rest.

If a heavy liquid be initially at rest at different levels in two communicating tubes, then motion will set in, so as to equalize the levels, for the centre of gravity of the system is thereby lowered, and the potential energy diminished. Equilibrium exists when the centre of gravity is at its lowest, and therefore the potential energy at a minimum, *i.e.* when the liquid stands at the same level in both tubes. If no special assumption be made with regard to the initial velocity of the liquid, the above proposition no longer holds. The potential energy need not decrease, and the higher level might rise or sink according to circumstances.

If our knowledge of thermal phenomena led us to recognize a state of minimum energy, a similar proposition would hold for this, but only for this, particular state. In reality no such minimum has been detected. It is, therefore, hopeless to seek to reduce the general laws regarding the direction of thermodynamical changes, as well as those of thermodynamical equilibrium, to the corresponding propositions in mechanics which hold good only for systems at rest.

§ **108.** Although these considerations make it evident that the principle of the conservation of energy cannot serve to determine the direction of a thermodynamical process, and therewith the conditions of thermodynamical equilibrium, unceasing attempts have been made to make the principle of the conservation of energy in some way or other serve this purpose. These attempts have, in many cases, stood in the way of a clear presentation of the second law. That attempts are still made to represent this law as contained in the principle of energy may be seen from the fact that the too restricted term " Energetics " is sometimes applied to all investigations on these questions. The conception of energy is not sufficient for the second law. It cannot be exhaustively treated by breaking up a natural process into a series of changes of energy, and then investigating the direction of each change. We can always tell, it is true, what are the different kinds of energy exchanged for one another; for there is no doubt that the principle of energy must be fulfilled, but the expression of the conditions of these changes remains arbitrary, and this ambiguity cannot be completely removed by any general assumption.

We often find the second law stated as follows : The change of mechanical work into heat may be complete, but, on the contrary, that of heat into work must needs be incomplete, since, whenever a certain quantity of heat is transformed into work, another quantity of heat must undergo a corresponding and compensating change; *e.g.* transference from higher to lower temperature. This is quite correct in certain very special cases, but it by no means expresses the essential feature of the process, as a simple example will show. An achievement which is closely associated with the discovery of the principle of energy, and which is one of the most important for the theory of heat, is the proposition expressed in equation (19), § 70, that the total internal energy of a gas depends only on the temperature, and not on the volume. If a perfect gas be allowed to expand, doing external work, and be prevented from cooling by connecting it with a heat-reservoir of higher temperature, the temperature of the gas, and at the same time its internal energy, remains unchanged, and it may be said that the amount of heat given out by the reservoir is completely changed into work without an exchange of energy taking place anywhere. Not the least objection can be made to this. The proposition of the " incomplete transformability of heat into work " cannot be applied to this case, except by a different way of viewing the process, which, however, changes nothing in the physical facts, and cannot, therefore, be confirmed or refuted by them, namely, by the introduction of new kinds of energy, only invented *ad hoc.* This consists in dividing the energy of the gas into several parts, which may then individually depend also on the volume. This division has, however, to be carried out differently for different cases (*e.g.* in one way for isothermal, in another for adiabatic processes), and necessitates complicated considerations even in cases of physical simplicity. But when we pass from the consideration of the first law of thermodynamics to that of the second, we have to deal with a new fact, and it is evident that no definition, however ingenious, although it contain no contradiction in itself, will ever permit of the deduction of a new fact.

§ **109.** There is but one way of clearly showing the

significance of the second law, and that is to base it on facts by formulating propositions which may be proved or disproved by experiment. The following proposition is of this character : It is in no way possible to completely reverse any process in which heat has been produced by friction. For the sake of example we shall refer to Joule's experiments on friction, described in § 60, for the determination of the mechanical equivalent of heat. Applied to these, our proposition says that, when the falling weights have generated heat in water or mercury by the friction of the paddles, no process can bo invented which will completely restore everywhere the initial state of that experiment, *i.e.* which will raise the weights to their original height, cool the liquid, and otherwise leave no change. The appliances used may be of any kind whatsoever, mechanical, thermal, chemical, electrical, etc., but the condition of *complete* restoration of the initial state renders it necessary that all materials and machines used must ultimately be left exactly in the condition in which they were before their application. Such a proposition cannot be proved *a priori*, neither does it amount to a definition, but it contains a definite assertion, to be stated precisely in each case, which may be verified by actual experiment. The proposition is therefore correct or incorrect.

§ **110.** Another proposition of this kind, and closely connected with the former, is the following : It is in no way possible to completely reverse any process in which a gas expands without performing work or absorbing heat, *i.e.* with constant total energy (as described in § 68). The word " completely " again refers to the accurate reproduction of the initial conditions. To test this, the gas, after it had assumed its new state of equilibrium, might first be compressed to its former volume by a weight falling to a lower level. External work is done on the gas, and it is thereby heated. The problem is now to bring the gas to its initial condition, and to raise the weight. The gas kept at constant volume might be reduced to its original temperature by conducting the heat of compression into a colder heat-reservoir. In order that the process may be completely reversed, the reservoir must be deprived of the heat gained thereby, and the weight raised to its original position. This is, however, exactly what was asserted in the preceding paragraph to be impracticable.

§ **111.** A third proposition in point refers to the conduction of heat. Supposing that a body receives a certain quantity of heat from another of higher temperature, the problem is to completely reverse this process, *i.e.* to convey back the heat without leaving any change whatsoever. In the description of Carnot's reversible cycle it has been pointed out, that heat can at any time be drawn from a heat-reservoir and transferred to a hotter reservoir without leaving any change except the expenditure of a certain amount of work, and the transference of an equivalent amount of heat from one reservoir to the other. If this heat could be removed, and the corresponding work recovered without other changes, the process of heat-conduction would be completely reversed. Here, again, we have the problem which was declared in § 109 to be impracticable.

Further examples of processes to which the same considerations apply are, diffusion, the freezing of an overcooled liquid, the condensation of a supersaturated vapour, all explosive reactions, and, in fact, every transformation of a system into a state of greater stability.

§ **112. Definition.**—A process which can in no way be completely reversed is termed *irreversible*, all other processes *reversible*. That a process may be irreversible, it is not sufficient that it cannot be directly reversed. This is the case with many mechanical processes which are not irreversible (*cf.* § 113). The full requirement is, that it be impossible, even with the assistance of all agents in nature, to restore everywhere the exact initial state when the process has once taken place. The propositions of the three preceding paragraphs, therefore, declare, that the generation of heat by friction, the expansion of a gas without the performance of external work and the absorption of external heat, the conduction of heat, etc., are irreversible processes.*

§ **113.** We now turn to the question of the actual existence of reversible and irreversible processes. Numerous reversible processes can at least be imagined, as, for instance; those consisting of a succession of states of equilibrium, as fully explained in § 71, and, therefore, directly reversible in all their parts. Further, all perfectly periodic processes, *e.g.* an ideal pendulum or planetary motion, are reversible, for, at the end of every period, the initial state is completely restored. Also, all mechanical processes with absolutely rigid bodies and absolutely incompressible liquids, as far as friction can be avoided, are reversible. By the introduction of suitable machines with absolutely unyielding connecting rods, frictionless joints and bearings, inextensible belts, etc., it is always possible to work the machines in such a way as to bring the system completely into its initial state without leaving any change in the machines, for the machines of themselves do not perform work.

If, for instance, a heavy liquid, originally at rest at different levels in two communicating tubes (§ 107), be set in motion by gravity, it will, in consequence of its kinetic energy, go beyond its position of equilibrium, and, since the tubes are supposed frictionless, again swing back to its exact original position. The process at this point has been completely reversed, and therefore belongs to the class of reversible processes.

As soon as friction is admitted, however, its reversibility is at least questionable. Whether reversible processes exist

---

* The principle that the conduction of heat is an irreversible process coincides exactly with the fundamental principle, which R. Clausius put in the forefront of his argument. The principle of Clausius states that *heat cannot of itself pass from a cold to a hot body.* As Clausius repeatedly and expressly pointed out, this principle does not merely say that heat does not flow directly from a cold to a hot body—that is self-evident, and is a condition of the definition of temperature—but it expressly states that heat can in no way and by no process be transported from a colder to a warmer body without leaving further changes, *i.e.* without *compensation.* Only in virtue of this wider meaning of the principle is it possible to draw conclusions about other natural processes.

in nature or not, is not *a priori* evident or demonstrable. There is, however, no purely logical objection to imagining that a means may some day be found of completely reversing some process hitherto considered irreversible : one, for example, in which friction or heat-conduction plays a part. But it can be demonstrated—and this will be done in the following chapter—that if, in a single instance, one of the processes declared to be irreversible in §§ 109, etc., should be found to be reversible, then all of these processes must be reversible in all cases. Consequently, either all or none of these processes are irreversible. There is no third possibility. If those processes are not irreversible, the entire edifice of the second law will crumble. None of the numerous relations deduced from it, however many may have been verified by experience, could then be considered as universally proved, and theoretical work would have to start from the beginning. (The so-called proofs of " energetics " are not a substitute, for a closer test shows all of them to be more or less imperfect paraphrases of the propositions to be proved. This is not the place, however, to demonstrate this point.) * It is this foundation on the physical fact of irreversibility which forms the strength of the second law. If, therefore, it must be admitted that a single experience contradicting that fact would render the law untenable, on the other hand, any confirmation of part supports the whole structure, and gives to deductions, even in seemingly remote regions, the full significance possessed by the law itself.

§ **114.** The significance of the second law of thermo-

---

* In many statements one finds the process of heat-conduction compared with the fall of a heavy liquid from a higher to a lower level. The law that in one case the heat falls from a higher to a lower temperature, and in the other the liquid from a higher to a lower level, is denoted as *the second fundamental principle of energetics.* This comparison shows very clearly the error of the case. Here no account is taken of the fact that the mechanical state of a body depends, not only on its position, but also on its velocity, while the thermal state depends only on the temperature. A heavy liquid may rise just as well as fall, while heat can only *fall.* In general, the *second law of energetics* is untrue. If the law is expressly limited to bodies at rest, it can be deduced, as has been shown in § 107, from the principle of energy, and it is therefore impossible to deduce anything new therefrom.

dynamics depends on the fact that it supplies a necessary and far-reaching criterion as to whether a definite process which occurs in nature is reversible or irreversible. Since the decision as to whether a particular process is irreversible or reversible depends only on whether the process can in any manner whatsoever be completely reversed or not, the nature of the initial and final states, and not the intermediate steps of the process, entirely settle it. The question is, whether or not it is possible, starting from the final state, to reach the initial one in any way without any other change. The second law, therefore, furnishes a relation between the quantities connected with the initial and final states of any natural process. The final state of an irreversible process is evidently in some way discriminate from the initial state, while in reversible processes the two states are in certain respects equivalent. The second law points out this characteristic property of both states, and also shows, when the two states are given, whether a transformation is possible in nature from the first to the second, or from the second to the first, without leaving changes in other bodies. For this purpose, of course, the two states must be fully characterized. Besides the chemical constitution of the systems in question, the physical conditions —viz. the state of aggregation, temperature, and pressure in both states—must be known, as is necessary for the application of the first law.

The relation furnished by the second law will evidently be simpler the nearer the two states are to one another. On this depends the great fertility of the second law in its treatment of cyclic processes, which, however complicated they may be, give rise to a final state only slightly different from the initial state. Since the system, which goes through the cyclic process, returns at the end to exactly the same state as at the beginning, we can leave it entirely out of account on comparing the two states (§ 91).

If we regard the second law from the mathematical point of view, the distinction between the final and initial states of a process can consist only in an inequality. This means that a certain quantity, which depends on the momentary state of the system, possesses in the final state a greater or smaller value, according to the definition of the sign of that quantity, than in the initial state.

The second law of thermodynamics states that there exists in nature for each system of bodies a quantity, which by all changes of the system either remains constant (in reversible processes) or increases in value (in irreversible processes). This quantity is called, following Clausius, the *entropy* of the system. The exposition in the following chapter aims at obtaining a mathematical expression for the entropy of a system, and proving its properties. First of all the entropy of ideal gases is obtained, since for these alone is an exact characteristic equation known, and therefrom also the entropies of all other substances.

§ **115.** Since there exists in nature no process entirely free from friction or heat-conduction, all processes which actually take place in nature, if the second law be correct, are in reality irreversible. Reversible processes form only an ideal limiting case. They are, however, of considerable importance for theoretical demonstration and for application to states of equilibrium.

CHAPTER II.

PROOF.

§ 116. SINCE the second fundamental principle of thermo-dynamics is, like the first, an empirical law, we can speak of its proof only in so far as its total purport may be deduced from a single simple law of experience about which there is no doubt. We, therefore, put forward the following proposition as being given directly by experience : *It is impossible to construct an engine which will work in a complete cycle, and produce no effect except the raising of a weight and the cooling of a heat-reservoir.**
Such an engine could be used simultaneously as a motor and a refrigerator without any waste of energy or material, and would in any case be the most profitable engine ever made. It would, it is true, not be equivalent to perpetual motion, for it does not produce work from nothing, but from the heat, which it draws from the reservoir. It would not, therefore, like perpetual motion, contradict the principle of energy, but would, nevertheless, possess for man the essential advantage of perpetual motion, the supply of work without cost; for the inexhaustible supply of heat in the earth, in the atmosphere, and in the sea, would, like the oxygen of the atmosphere, be at everybody's immediate disposal. For this reason we take the above proposition as our starting point. Since we are to deduce the second law from it, we expect, at the same time, to make a most serviceable application of any natural phenomenon which may be discovered to deviate from the second law. As soon as a phenomenon is found to con-tradict any legitimate conclusions from the second law, this

* The temperature of the reservoir does not enter into the question. If such a machine is possible with a reservoir at 1000° C. it is also possible with a reservoir at 0° C. To see this one has only to use a suitably devised Carnot cycle (§ 91).

contradiction must arise from an inaccuracy in our first assumption, and the phenomenon could be used for the construction of the above-described engine. We shall in the following, according to the proposal of Ostwald, speak of perpetual motion of the second kind, since it stands in the same relation to the second law as perpetual motion of the first kind does to the first law. In connection with all objections to the second law, it must be borne in mind that, if no errors are to be found in the line of proof, they are ultimately directed against the impossibility of perpetual motion of the second kind (§ 136).*

§ **117.** From the impossibility of perpetual motion of the second kind, it follows, in the first place, that the generation of heat by friction is *irreversible* (*cf.* def. § 112). For supposing it were not so, *i.e.* supposing a method could be found by which a process involving generation of heat by friction could be completely reversed, this very method would produce what is identically perpetual motion of the second kind : viz. a change which consists of nothing but the production of work, and the absorption of an equivalent amount of heat.

§ **118.** It follows, further, that the expansion of a gas without the performance of external work, or the absorption of heat, is irreversible. For, suppose a method were known of completely reversing this process, *i.e.* of reducing the volume of a gas, without leaving any other change whatsoever, this method could be utilized for the production of perpetual motion of the second kind in the following manner. Allow the gas to do work by expansion, supplying the energy

---

* The starting point selected by me for the proof of the second law coincides fundamentally with that which R. Clausius, or which Sir W. Thomson, or which J. Clerk Maxwell used for the same purpose. The fundamental proposition which each of these investigators placed at the beginning of his deductions asserts each time, only in different form, the impossibility of the realization of perpetual motion of the second kind. I have selected the above form of expression, because of its evident technical significance. Not a single really rational proof of the second law has thus far been advanced which does not require this fundamental principle, however numerous the attempts in this direction may have been, nor do I believe that such an attempt will ever meet with success.

lost thereby by the conduction of heat from a reservoir at the same or higher temperature, and then, by the assumed method, reduce the volume of the gas to its initial value without leaving any other change. This process might be repeated as often as we please, and would therefore represent an engine working in a complete cycle, and producing no effect except the performance of work, and the withdrawal of heat from a reservoir, *i.e.* perpetual motion of the second kind.

On the basis of the proposition we have just proved, that the expansion of a gas without the performance of work and the absorption of heat is irreversible, we shall now carry through the proof of the second law for those bodies whose thermodynamical properties are most completely known, viz. for perfect gases.

§ **119.** If a perfect gas be subjected to infinitely slow compression or expansion, and if, at the same time, heat be applied or withdrawn, we have, by equation (22), in each infinitely small portion of the process, per unit mass,

$$q = du + pdv$$

or, since for a perfect gas,

$$du = c_v d\mathrm{T},$$

and

$$p = \frac{\mathrm{R}}{m} \cdot \frac{\mathrm{T}}{v},$$

$$q = c_v d\mathrm{T} + \frac{\mathrm{R}}{m} \cdot \frac{\mathrm{T}}{v} dv.$$

If the process be adiabatic, then $q = 0$, and the integration of the above equation gives (as in § 88) the function

$$c_v \log \mathrm{T} + \frac{\mathrm{R}}{m} \log v$$

equal to a constant. We shall now put

$$\phi = c_v \log \mathrm{T} + \frac{\mathrm{R}}{m} \log v + \text{const.,} \quad . \quad . \quad (51)$$

and call this function, after Clausius, the *entropy* of unit mass of the gas. The constant, which has to be added, can be determined by arbitrarily fixing the zero state. Accordingly

$$\Phi = M\phi = M\left(c_v \log T + \frac{R}{m} \log v + \text{const.}\right) \quad . \quad (52)$$

is the entropy of mass M of the gas. The entropy of the gas, therefore, remains constant during the described adiabatic change of state.

§ **120.** On the application of heat, the entropy of the gas changes, in the case considered, by

$$d\Phi = M\left(c_v \frac{dT}{T} + \frac{R}{m} \cdot \frac{dv}{v}\right) = M\frac{du + pdv}{T} \quad . \quad (53)$$

$$d\Phi = M\frac{q}{T} = \frac{Q}{T} \quad . \quad . \quad . \quad . \quad . \quad . \quad (53a)$$

It increases or decreases according as heat is absorbed or evolved.

The absorbed heat Q has here been broken up into two factors, T and $d\Phi$. According to a view which has recently been brought forward, this breaking up of heat into factors is regarded as a general property of heat. It should, however, be emphasized that equation (53a) is by no means generally true. It holds only in the particular case where the external work performed by the gas is expressed by pdV. The relation (53) holds, quite generally, for any process in which the temperature of the gas is increased by dT, and the volume by dV. It is, in fact, only a different mathematical form for the definition of the entropy given in (52). It holds, for example, also when a gas, as in the process described in § 68, passes, without the performance of external work, into a new state of equilibrium at the same temperature with greater volume. On the other hand, the equation

$$Q = dU + pdV$$

holds by no means in all cases, but should, in general, be replaced by

$$Q + W = dU,$$

where W, the work done on the substance, may have any value within certain limits. For instance, $W = 0$, if the gas expands without performing external work. In this case, $Q = dU$, and the equation $Q = Td\Phi$ no longer holds.

§ **121.** We shall now consider two gases which can communicate heat to one another by conduction, but may, in general, be under different pressures. If the volume of one, or both, of the gases be changed by some reversible process, care being taken that the temperatures of the gases equalize at each moment, and that no exchange of heat takes place with surrounding bodies, we have, according to equation (53), during any element of time, for the first gas,

$$d\Phi_1 = \frac{Q_1}{T_1},$$

and, for the second gas,

$$d\Phi_2 = \frac{Q_2}{T_2}.$$

According to the conditions of the process,

$$T_1 = T_2 \text{ and } Q_1 + Q_2 = 0,$$

whence, $\qquad d\Phi_1 + d\Phi_2 = 0$

or, for a finite change,

$$\Phi_1 + \Phi_2 = \text{const.} \quad . \quad . \quad . \quad . \quad (54)$$

The sum of the entropies of the two gases remains constant during the described process.

§ **122.** Any such process with two gases is evidently reversible in all its parts, for it may be directly reversed without leaving changes in the surroundings. From this follows the proposition that it is always possible to bring two gases, by a reversible process, without leaving changes in other bodies, from any given state to any other given state, if the sum of the entropies in the two states be equal.*

---

* Since the states of both gases are arbitrary, the gas will in general not have the same total energy in the one state as in the other. The passage of the gas from one state to another is accompanied by a loss or gain of energy. This, however, introduces no difficulty, for we can always regard the energy

Let an initial state of the gases be given by the temperatures T_1, T_2, and the specific volumes v_1, v_2; a second state by the corresponding values T_1', T_2'; v_1', v_2'. We now suppose that

$$\Phi_1 + \Phi_2 = \Phi_1' + \Phi_2' \quad \cdots \quad (55)$$

Bring the first gas to the temperature T_2 by a reversible adiabatic compression or expansion; then place the two gases in thermal contact with one another, and continue to compress or expand the first infinitely slowly. Heat will now pass between the two gases, and the entropy of the first one will change, and it will be possible to make this entropy assume the value Φ_1'. But, according to (54), during the above process the sum of the two entropies remains constant, and $= \Phi_1 + \Phi_2$; therefore the entropy of the second gas is $(\Phi_1 + \Phi_2) - \Phi_1'$, which is, according to (55), equal to Φ_2'. If we now separate the two gases, and compress or expand each one adiabatically and reversibly until they have the required temperatures T_1' and T_2', the specific volumes must then be v_1' and v_2', and the required final state has been reached.

This process is reversible in all its parts, and no changes remain in other bodies; * in particular, the surroundings have neither gained nor lost heat.† The conditions of the problem have therefore been fulfilled, and the proposition proved.

§ **123.** A similar proposition can readily be proved for any number of gases. It is always possible to bring a system of n gases from any one state to any other by a reversible process without leaving changes in other bodies, if the sum of the entropies of all the gases is the same in both states, *i.e.* if

$$\Phi_1 + \Phi_2 + \ldots + \Phi_n = \Phi_1' + \Phi_2' + \ldots + \Phi_n'. \quad (56)$$

By the process described in the preceding paragraph we may,

---

as mechanical work, produced by the raising or lowering of weights. Since the weights change only their position and not their internal state, no changes are left in them, since in general all purely mechanical processes are essentially reversible (§ 113).

* The necessary work can be performed by the raising or lowering of unchangeable weights involving no internal changes.

† Also there have been no density changes, for we can imagine the gas holder in a vacuum.

by the successive combination of pairs of gases of the system, bring the first, then the second, then the third, and so on to the $(n-1)$th gas, to the required entropy. Now, in each of the successive processes the sum of the entropies of all the gases remains constant, and, since the entropies of the first $(n-1)$ gases are Φ_1', Φ_2' ... Φ'_{n-1}, the entropy of the nth gas is necessarily

$$(\Phi_1 + \Phi_2 + \ldots + \Phi_n) - (\Phi_1' + \Phi_2' + \ldots + \Phi'_{n-1}).$$

This is, according to (56), the required value Φ_n'. Each gas can now be brought by an adiabatic reversible process into the required state, and the problem is solved.

If we call the sum of the entropies of all the gases the entropy of the whole system, we may then say : *If a system of gases has the same entropy in two different states, it may be transformed from the one to the other by a reversible process, without leaving changes in other bodies.*

§ **124.** We now introduce the proposition proved in § 118, that the expansion of a perfect gas, without performing external work or absorbing heat, is irreversible; or, what is the same thing, that the transition of a perfect gas to a state of greater volume and equal temperature, without external effects, as described in § 68, is irreversible. Such a process corresponds to an increase of the entropy, according to the definition (52). It immediately follows *that it is altogether impossible to decrease the entropy of a gas without producing changes in other bodies.* If this were possible, the irreversible expansion of a gas could be completely reversed. After the gas had expanded without external effects, and had assumed its new state of equilibrium, the entropy of the gas could be reduced to its initial value, without leaving changes in other bodies, by the supposed method, and then, by an adiabatic reversible process, during which the entropy of the gas remains constant, brought to its initial temperature, and thereby also to its original volume. This would completely reverse * the

---

* That no mechanical work remains over follows from the first law of thermodynamics, since with the original state of the gas also the original energy is restored.

first expansion, and furnish, according to § 118, perpetual motion of the second kind.

§ **125.** A system of two or more gases behaves in the same way. There exists, in nature, no means of diminishing the entropy of a system of perfect gases, without leaving changes in bodies outside the system. A contrivance which would accomplish this, be it mechanical, thermal, chemical, or electrical in nature, might be used to reduce the entropy of a single gas without leaving changes in other bodies.

Suppose a system of gases to have passed in any manner from one state in which their entropies are $\Phi_1, \Phi_2 \ldots \Phi_n$, to a state where they are $\Phi_1', \Phi_2' \ldots \Phi_n'$, and that no change has been produced in any body outside the system, and let

$$\Phi_1' + \Phi_2' + \ldots + \Phi_n' < \Phi_1 + \Phi_2 + \ldots + \Phi_n, \quad (57)$$

then it is possible, according to the proposition proved in § 123, to bring the system by a reversible process, without leaving changes in other bodies, into any other state in which the sum of the entropies is

$$\Phi_1' + \Phi_2' + \ldots + \Phi_n',$$

and accordingly into a state in which the first gas has the entropy Φ_1, the second the entropy $\Phi_2 \ldots$, the $(n-1)$th the entropy Φ_{n-1}, and the nth in consequence the entropy

$$(\Phi_1' + \Phi_2' + \ldots + \Phi_n') - \Phi_1 - \Phi_2 - \ldots - \Phi_{n-1} \quad (58)$$

The first $(n-1)$ gases may now be reduced to their original state by reversible adiabatic processes. The nth gas possesses the entropy (58), which is, according to the supposition (57), smaller than the original entropy Φ_n. The entropy of the nth gas has, therefore, been diminished without leaving changes in other bodies.* This we have already proved in the preceding paragraph to be impossible.

The general proposition has, therefore, been proved, and we may immediately add the following.

---

* Raising or lowering of weights are not internal changes; see the note to § 122.

§ **126.** *If a system of perfect gases pass in any way from one state to another, and no changes remain in surrounding bodies, the entropy of the system is certainly not smaller, but either greater than, or, in the limit, equal to that of the initial state ; in other words, the total change of the entropy ≥ 0.* The sign of inequality corresponds to an irreversible process, the sign of equality to a reversible one. The equality of the entropies in both states is, therefore, not only a sufficient, as described in § 123, but also a necessary condition of the complete reversibility of the transformation from the one state to the other, provided no changes are to remain in other bodies.

§ **127.** The scope of this proposition is considerable, since there have designedly been imposed no restrictions regarding the way in which the system passes from its initial to its final state. The proposition, therefore, holds not only for slow and simple processes, but also for physical and chemical ones of any degree of complication, provided that at the end of the process no changes remain in any body outside the system. It must not be supposed that the entropy of a gas has a meaning only for states of equilibrium. We may assume each sufficiently small particle, even of a gas in turmoil, to be homogeneous and at a definite temperature, and must, therefore, according to (52), assign to it a definite value of the entropy. M, v, and T are then the mass, specific volume, and temperature of the particle under consideration. A summation extending over all the particles of the mass—within which the values of v and T may vary from particle to particle—gives the entropy of the whole mass of the gas in the particular state. The proposition still holds, that the entropy of the whole gas must continually increase during any process which does not give rise to changes in other bodies, *e.g.* when a gas flows from a vessel into a vacuum (§ 68). It will be seen that the velocity of the gas particles does not influence the value of the entropy [*]; neither does their height above a certain horizontal plane, although they are considered to have weight.

[*] If the motion of the gas is so tumultuous that neither the temperature nor the density can be defined, naturally the definition of entropy loses its

§ **128.** The laws which we have deduced for perfect gases may be transferred to any substance in exactly the same way. The main difference is, that the expression for the entropy of any body cannot, in general, be written down in finite quantities, since the characteristic equation is not generally known. But it can be demonstrated—and this is the deciding point—that, for any other body, there exists a function with the characteristic properties of the entropy.

Imagine any homogeneous body to pass through a certain reversible or irreversible cycle and to be brought back to its exact original state, and let the external effects of this process consist in the performance of work and in the addition or withdrawal of heat. The latter may be brought about by means of any required number of suitable heat-reservoirs. After the process, no changes remain in the substance itself; the heat-reservoirs alone have suffered change. We shall now assume all the heat-reservoirs to be perfect gases, kept either at constant volume or at constant pressure, but, at any rate, subject only to reversible changes of volume.* According to our last proposition, the sum of the entropies of all these gases cannot have decreased, since after the process no change remains in any other body.

If Q denote the amount of heat given to the substance during an infinitely small element of time by one of the reservoirs; T, the temperature of the reservoir at that moment; † then, according to equation (53a), the reservoir's change of entropy during that element of time is

$$- \frac{Q}{T}.$$

The change of the entropy of all the reservoirs, during all the elements of time considered, is

---

meaning. For this case, as L. Boltzmann has shown, another definition can be given from the standpoint of the kinetic theory of gases. This definition possesses a still more general meaning, and passes into the usual one, when the states are stationary or nearly stationary.

* The supposition that the gases are ideal is not dependent on any limit of temperature. For, also at the lowest temperatures, each gas behaves sensibly as a perfect gas, if the density is taken sufficiently small.

† The simultaneous temperature of the body is here of no importance.

$$- \Sigma \frac{Q}{T}.$$

Now, according to § 126, we have the following condition :

$$- \Sigma \frac{Q}{T} \geqq 0$$

or
$$\Sigma \frac{Q}{T} \leqq 0.$$

This is the form in which the second law was first enunciated by Clausius.

A further condition is given by the first law; for, according to (17) in § 63, we have, during every element of time of the process,

$$Q + W = dU,$$

where U is the energy of the body, and W the work done on the body during the element of time.

§ **129.** If we now make the special assumption that the external pressure is, at any moment, equal to the pressure p of the substance, the work of compression becomes, according to (20),

$$W = - pdV,$$
whence $$Q = dU + pdV.$$

If, further, each heat-reservoir be exactly at the temperature of the substance at the moment when brought into operation, the cyclic process is reversible, and the inequality of the second law becomes an equality :

$$\Sigma \frac{Q}{T} = 0,$$

or, on substituting the value of Q,

$$\Sigma \frac{dU + pdV}{T} = 0.$$

All the quantities in this equation refer to the state of the substance itself. It admits of interpretation without reference to the heat-reservoirs, and amounts to the following proposition.

§ 130. *If a homogeneous body be taken through a series of states of equilibrium (§ 71), that follow continuously from one another, back to its initial state, then the summation of the differential*

$$\frac{d\mathrm{U} + pd\mathrm{V}}{\mathrm{T}}$$

extending over all the states of that process gives the value zero. It follows that, if the process be not continued until the initial state, 1, is again reached, but be stopped at a certain state, 2, the value of the summation

$$\int_{1}^{2}\frac{d\mathrm{U} + pd\mathrm{V}}{\mathrm{T}} \quad . \quad . \quad . \quad . \quad . \quad (59)$$

depends only on the states 1 and 2, not on the manner of the transformation from state 1 to state 2. If two series of changes leading from 1 to 2 be considered (*e.g.* curves α and β in Fig. 2, § 75), these can be combined into an infinitely slow cyclic process. We may, for example, go from 1 to 2 along α, and return to 1 along β.

It has been demonstrated that over the entire cycle :

$$\int_{1}^{2}\!\!{}_{(a)}\frac{d\mathrm{U} + pd\mathrm{V}}{\mathrm{T}} + \int_{2}^{1}\!\!{}_{(\beta)}\frac{d\mathrm{U} + pd\mathrm{V}}{\mathrm{T}} = 0,$$

whence $\qquad\displaystyle\int_{1}^{2}\!\!{}_{(a)}\frac{d\mathrm{U} + pd\mathrm{V}}{\mathrm{T}} = \int_{1}^{2}\!\!{}_{(\beta)}\frac{d\mathrm{U} + pd\mathrm{V}}{\mathrm{T}}.$

The integral (59) with the above-proved properties has been called by Clausius the *entropy* of the body in state 2, referred to state 1 as the zero state. The entropy of a body in a given state, like the internal energy, is completely determined except for an additive constant, whose value depends on the zero state.

Denoting the entropy, as formerly, by Φ, we have :

$$\Phi = \int \frac{dU + pdV}{T}$$

and

$$d\Phi = \frac{dU + pdV}{T} . \quad . \quad . \quad . \quad (60)$$

or per unit mass :

$$d\phi = \frac{du + pdv}{T} \quad . \quad . \quad . \quad . \quad (61)$$

This, again, leads to the value (51) for a perfect gas. The expression for the entropy of any body may be found by immediate integration (§ 254), provided its energy, $U = Mu$, and its volume, $V = Mv$, are known as functions, say, of T and p. Since, however, these are not completely known except for perfect gases, we have to content ourselves in general with the differential equation. For the proof, and for many applications of the second law, it is, however, sufficient to know that this differential equation contains in reality a unique definition of the entropy.

§ **131.** We may, therefore, just as in the case of perfect gases, speak of the entropy of any substance as of a finite quantity determined by the momentary values of temperature and volume, even when the substance undergoes reversible or irreversible changes. The differential equation (61) holds, as was stated in § 120 in the case of perfect gases, for any change of state, including irreversible changes. This more general application of the conception of the entropy in no wise contradicts the manner of its deduction. The entropy of a body in any given state is measured by means of a reversible process which brings the body from that state to the zero state. This ideal process, however, has nothing to do with any actual reversible or irreversible changes which the body may have undergone or be about to undergo.

On the other hand, it should be stated that the differential equation (60), while it holds for changes of volume and temperature, does not apply to changes of mass, for this kind of change was in no way referred to in the definition of the entropy.

Finally, we shall call the sum of the entropies of a number of bodies briefly the entropy of the system composed of those bodies. Thus the entropy of a body whose particles are not at uniform temperature, and have different velocities, may be found, as in the case of gases (§ 127), by a summation extending over all its elements of mass, provided the temperature and density within each infinitely small element of mass may be considered uniform. Neither the velocity nor the weight of the particles enter into the expression for the entropy.

§ **132.** The existence and the value of the entropy having been established for all states of a body, there is no difficulty in transferring the proof, which was given for perfect gases (beginning in § 119), to any system of bodies. Just as in § 119 we find that, during reversible adiabatic expansion or compression of a body, its entropy remains constant, while by the absorption of heat the change of the entropy is

$$d\Phi = \frac{Q}{T} \quad . \quad . \quad . \quad . \quad . \quad (62)$$

This relation holds only for reversible changes of volume, as was shown for perfect gases in § 120. Besides, it is found, as in § 121, that during reversible expansion or compression of two bodies at a common temperature, if they be allowed to exchange heat by conduction with one another, but not with surrounding bodies, the sum of their entropies remains constant. A line of argument corresponding fully to that advanced for perfect gases then leads to the following general result : * *It is impossible in any way to diminish the entropy of a system of bodies without thereby leaving behind changes in other bodies.* If, therefore, a system of bodies has changed its state in a physical or chemical way, without leaving any change in bodies not belonging to the system, then the entropy in the

---

* The generalization of the theorem, which was proved in § 124 for a perfect gas, presents a certain difficulty when the body is incompressible. The volume cannot be altered, and, accordingly, no irreversible expansion is possible. The proof can, however, be easily extended with the help of a perfect gas, which can be brought into heat communication with the body. The entropy of the body can thereby be suitably altered.

final state is greater than, or, in the limit, equal to the entropy in the initial state. The limiting case corresponds to reversible, all others to irreversible, processes.

§ **133.** The restriction, hitherto indispensable, that no changes must remain in bodies outside the system is easily dispensed with by including in the system all bodies that may be affected in any way by the process considered. The proposition then becomes : *Every physical or chemical process in nature takes place in such a way as to increase the sum of the entropies of all the bodies taking any part in the process. In the limit, i.e. for reversible processes, the sum of the entropies remains unchanged.* This is the most general statement of the second law of Thermodynamics.

§ **134.** As the impossibility of perpetual motion of the first kind leads to the first law of Thermodynamics, or the principle of the conservation of energy; so the impossibility of perpetual motion of the second kind has led to the second law, properly designated as the *principle of the increase of the entropy.** This principle may be presented under other forms, which possess certain practical advantages, especially for isothermal or isobaric processes. They will be mentioned in our next chapter. It should be emphasized, however, that the form here given is the only one of unrestricted applicability to any finite process, and that no other universal measure of the irreversibility of processes exists than the amount of the increase of the entropy to which they lead. All other forms of the second law are either applicable to infinitesimal changes only, or presuppose, when extended to finite changes, the existence of some special condition imposed upon the process (§§ 140, etc.).

The real meaning of the second law has frequently been

---

* That the first law is expressed by an equality, but the second by an inequality naturally depends on the fact that the law of the impossibility of perpetual motion of the first kind is also reversible, *i.e.* work can neither be absolutely created nor absolutely destroyed, while, on the other hand, the law of the impossibility of perpetual motion of the second kind permits no reversibility, since it is quite possible to construct a machine which does nothing but use up work and warm a reservoir correspondingly.

looked for in a " dissipation of energy." This view, proceeding, as it does, from the irreversible phenomena of conduction and radiation of heat, presents only one side of the question. There are irreversible processes in which the final and initial states show exactly the same form of energy, *e.g.* the diffusion of two perfect gases (§ 238), or further dilution of a dilute solution. Such processes are accompanied by no perceptible transference of heat, nor by external work, nor by any noticeable transformation of energy.* They occur only for the reason that they lead to an appreciable increase of the entropy.† The amount of " lost work " yields a no more definite general measure of irreversibility than does that of " dissipated energy." This is possible only in the case of isothermal processes (§ 143).

§ **135.** Clausius summed up the first law by saying that the energy of the world remains constant; the second by saying that the entropy of the world tends towards a maximum. Objection has justly been raised to this form of expression. The energy and the entropy of the world have no meaning, because such quantities admit of no accurate definition. Nevertheless, it is not difficult to express the characteristic feature of those propositions of Clausius in such a way as to give them a meaning, and to bring out more clearly what Clausius evidently wished to express by them.

The energy of any system of bodies changes according to the measure of the effects produced by external agents. It remains constant, only, if the system be isolated. Since, strictly speaking, every system is acted on by external agents— for complete isolation cannot be realized in nature—the energy of a finite system may be approximately, but never absolutely, constant. Nevertheless, the more extended the system, the more negligible, in general, will the external effects become, in comparison with the magnitude of the energy of the system, and the changes of energy of its parts (§ 66); for, while the

---

* At any rate, if we stick to the definition given in § 56 and not introduce new *ad hoc* kinds of energy.

† In this case it would be more to the point to speak of a dissipation of matter than of a dissipation of energy.

external effects are of the order of magnitude of the surface of the system, the internal energy is of the order of magnitude of the volume. In very small systems (elements of volume) the opposite is the case for the same reason, since here the energy of the system may be neglected in comparison with any one of the external effects. Frequent use is made of this proposition, *e.g.* in establishing the limiting conditions in the theory of the conduction of heat.

In the case here considered, it may, therefore, be said that the more widely extended a system we assume, the more approximately, in general, will its energy remain constant.* A comparatively small error will be committed in assuming the energy of our solar system to be constant, a proportionately smaller one if the system of all known fixed stars be included. In this sense an actual significance belongs to the proposition, that the energy of an infinite system, or the energy of the world, remains constant.†

The proposition regarding the increase of the entropy should be similarly understood. If we say that the entropy of a system increases quite regardless of all outside changes, an error will, in general, be committed, but the more comprehensive the system, the smaller does the proportional error become.

§ **136.** In conclusion, we shall briefly discuss the question of the possible limitations to the second law.‡ If there exist any such limitations—a view still held by many scientists and philosophers—this much may be asserted, that their existence

---

* This proposition holds quite generally for all physical processes, if action at a distance be excluded.

† This may be expressed analytically as follows. Let E denote the total energy contained in a very large space S, then we have the equation :

$$lt \frac{1}{E} \frac{dE}{dt} = 0,$$
$$S = \infty.$$

The change in the quantity log E with the time is smaller the greater S becomes, and approaches indefinitely near to the constant.

‡ The following discussion, of course, deals with the meaning of the second law only in so far as it can be surveyed from the points of view contained in this work avoiding all atomic hypotheses.

presupposes an error in our starting-point, viz. the impossibility of perpetual motion of the second kind, or a fault in our
method of proof. From the beginning we have recognized the
legitimacy of the first of these objections, and it cannot be
removed by any line of argument. The second objection
generally amounts to the following. The impracticability of
perpetual motion of the second kind is granted, yet its absolute
impossibility is contested, since our limited experimental
appliances, supposing it were possible, would be insufficient for
the realization of the ideal processes which the line of proof
presupposes. This position, however, proves untenable. It
would be absurd to assume that the validity of the second
law depends in any way on the skill of the physicist or chemist
in observing or experimenting. The gist of the second law
has nothing to do with experiment; the law asserts briefly
that *there exists in nature a quantity which changes always in
the same sense in all natural processes.* The proposition stated
in this general form may be correct or incorrect; but which-
ever it may be, it will remain so, irrespective of whether
thinking and measuring beings exist on the earth or not, and
whether or not, assuming they do exist, they are able to
measure the details of physical or chemical processes more
accurately by one, two, or a hundred decimal places than we
can. The limitations to the law, if any, must lie in the same
province as its essential idea, in the observed Nature, and not
in the Observer. That man's experience is called upon in the
deduction of the law is of no consequence; for that is, in fact,
our only way of arriving at a knowledge of natural law. But
the law once discovered must receive recognition of its independence, at least in so far as Natural Law can be said to exist
independent of Mind. Whoever denies this must deny the
possibility of natural science.

The case of the first law is quite similar. To most un-
prejudiced scientists the impossibility of perpetual motion
of the first kind is certainly the most direct of the general
proofs of the principle of energy. Nevertheless, hardly any
one would now think of making the validity of that principle

depend on the degree of accuracy of the experimental proof of that general empirical proposition. Presumably the time will come when the principle of the increase of the entropy will be presented without any connection with experiment. Some metaphysicians may even put it forward as being *a priori* valid. In the mean time, no more effective weapon can be used by both champions and opponents of the second law than indefatigable endeavour to follow the real purport of this law to the utmost consequences, taking the latter one by one to the highest court of appeal—experience. Whatever the decision may be, lasting gain will accrue to us from such a proceeding, since thereby we serve the chief end of natural science—the enlargement of our stock of knowledge.

CHAPTER III.

GENERAL DEDUCTIONS.

§ 137. Our first application of the principle of the entropy which was expressed in its most general form in the preceding chapter, will be to Carnot's cycle, described in detail for perfect gases in § 90. This time, the system operated upon may be of any character whatsoever, and chemical reactions, too, may take place, provided they are reversible. Resuming the notation used in § 90, we may at once state the result.

In a cyclic process, according to the first law, the heat, Q_2, given out by the hotter reservoir is equivalent to the sum of the work done by the system, $W' = - W$, and the heat received by the colder reservoir, $Q_1' = - Q_1$:

$$Q_2 = W' + Q_1'$$

or
$$Q_1 + Q_2 + W = 0 \quad . \quad . \quad . \quad . \quad (63)$$

According to the second law, since the process is reversible, all bodies which show any change of state after the process, *i.e.* the two heat-reservoirs only, possess the same total entropy as before the process. The change of the entropy of the two reservoirs is, according to (62) :

$$\frac{Q_1'}{T_1} = - \frac{Q_1}{T_1} \text{ for the first, and } - \frac{Q_2}{T_2} \text{ for the second,} \quad (64)$$

their sum :
$$\frac{Q_1}{T_1} + \frac{Q_2}{T_2} = 0 \quad . \quad . \quad . \quad . \quad (65)$$

whence, by (63),

$$Q_1 : Q_2 : W = (- T_1) : T_2 : (T_1 - T_2)$$

as in (44), but without any assumption as to the nature of the substance passing through the cycle of operations.

In order, therefore, to gain the mechanical work, W',

by means of a reversible Carnot cycle of operations with any substance between two heat-reservoirs at the temperatures T_1 and T_2 ($T_2 > T_1$), the quantity of heat

$$Q_1' = \frac{T_1}{T_2 - T_1} W'$$

must pass from the hotter to the colder reservoir. In other words, the passage of the quantity of heat Q_1' from T_2 to T_1 may be taken advantage of to gain the mechanical work

$$W' = \frac{T_2 - T_1}{T_1} Q_1' \quad . \quad . \quad . \quad (66)$$

§ 138. For an irreversible cycle, *i.e.* one involving any irreversible physical or chemical changes of the substance operated upon, the equation of energy (63) still holds, but the equation for the change of the entropy (65) is replaced by the inequality :

$$-\frac{Q_1}{T_1} - \frac{Q_2}{T_2} > 0.$$

Observe, however, that the expressions (64) for the change of the entropy of the reservoirs are still correct, provided we assume that T_1 and T_2 denote the temperatures of the reservoirs, and that any changes of volume of the substances used as reservoirs are reversible. Thus,

$$\frac{Q_1}{T_1} + \frac{Q_2}{T_2} < 0 \quad . \quad . \quad . \quad . \quad (67)$$

or

$$Q_2 < \frac{T_2}{T_1} Q_1',$$

hence, from (67) and (63),

$$W' = -W = Q_1 + Q_2 < Q_1 + \frac{T_2}{T_1} Q_1'$$

or

$$W' < \frac{T_2 - T_1}{T_1} Q_1'.$$

This means that the amount of work, W', to be gained by means of a cyclic process from the transference of the heat,

Q_1', from a hotter to a colder reservoir, is always smaller for an irreversible process than for a reversible one. Consequently the equation (66) represents the maximum amount of work to be gained from a cyclic process with any system by the passage of heat Q_1' from a higher temperature T_2 to a lower temperature T_1.

In particular, if $W' = 0$, it follows from the equation of energy (63) that

$$Q_2 = - Q_1 = Q_1'$$

and the inequality (67) becomes

$$Q_2\left(\frac{1}{T_2} - \frac{1}{T_1}\right) < 0.$$

In this case the cyclic process results in the transference of heat (Q_2) from the reservoir of temperature T_2 to that of temperature T_1, and the inequality means that this flow of heat is always directed from the hotter to the colder reservoir.

Again, a special case of this type of process is the direct passage of heat by conduction between heat-reservoirs, without any actual participation of the system supposed to pass through the cycle of operations. It is seen to be an irreversible change, since it brings about an increase of the sum of the entropies of the two heat-reservoirs.

§ **139.** We shall now apply the principle of the entropy to any reversible or irreversible cycle with any system of bodies, in the course of which only one heat-reservoir of constant temperature T is used. Whatever may be the nature of the process in detail, there remains at its close no change of the entropy except that undergone by the heat-reservoir. According to the first law, we have

$$W + Q = 0.$$

W is the work done on the system, and Q the heat absorbed by the system from the reservoir.

According to the second law, the change of the entropy

of the reservoir, within which only reversible changes of volume are supposed to take place, is

$$-\frac{Q}{T} \geqq 0$$

or
$$Q \leqq 0,$$

whence
$$W \geqq 0.$$

Work has been expended on the system, and heat added to the reservoir. This inequality formulates analytically the impossibility of perpetual motion of the second kind.

If, in the limit, the process be reversible, the signs of inequality disappear, and *both the work* W *and the heat* Q *are zero.* On this proposition rests the great fertility of the second law in its application to isothermal reversible cycles.

§ **140.** We shall no longer deal with cycles, but shall consider the general question of the direction in which a change will set in, when any system in nature is given. For chemical reactions in particular is this question of importance. It is completely answered by the second law in conjunction with the first, for the second law contains a condition necessary for all natural processes. Let us imagine any homogeneous or heterogeneous system of bodies at the common temperature T, and investigate the conditions for the starting of any physical or chemical change.

According to the first law, we have for any infinitesimal change :

$$d\mathrm{U} = Q + W, \quad \cdots \cdots \quad (68)$$

where U is the total internal energy of the system, Q the heat absorbed by the system during the process, and W the work done on the system.

According to the second law, the change of the total entropy of all the bodies taking part in the process is

$$d\Phi + d\Phi_0 \geqq 0$$

where Φ is the entropy of the system, Φ_0 the entropy of the surrounding medium (air, calorimetric liquid, walls of vessels, etc.). Here the sign of equality holds for reversible cases, which, it is true, should be considered as an ideal limiting case of actual processes (§ 115).

If we assume that all changes of volume in the surrounding medium are reversible, we have, according to (62),

$$d\Phi_0 = -\frac{Q}{T},$$

also

$$d\Phi - \frac{Q}{T} \geq 0$$

or, by (68),

$$d\Phi - \frac{dU - W}{T} \geq 0 \quad . \quad . \quad . \quad (69)$$

or

$$dU - Td\Phi \leq W \quad . \quad . \quad . \quad (70)$$

All conclusions with regard to thermodynamic chemical changes, hitherto drawn by different authors in different ways, culminate in this relation (70). It cannot in general be integrated, since the left-hand side is not, in general, a perfect differential. The second law, then, does not lead to a general statement with regard to finite changes of a system taken by itself unless something be known of the external conditions to which it is subject. This was to be expected, and holds for the first law as well. To arrive at a law governing finite changes of the system, the knowledge of such external conditions as will permit the integration of the differential is indispensable.* Among these the following are singled out as worthy of note.

§ **141. Case I. Adiabatic Process.**—No exchange of heat with the surroundings being permitted, we have $Q = 0$, and, by (68),

$$dU = W.$$

Consequently, by (70), $d\Phi \geq 0$.

---

* One often speaks of an *isolated* system, but this, of course, is not to be taken as if in certain cases definite changes can take place in the system without prescribing any external conditions. In nature, there can be no finite isolated system. Some surface conditions are always present. It may be that the system is contained in a vessel with solid or elastic walls, or that it is bounded by the free atmosphere, or that it is contained in a vacuum, etc.

The entropy of the system increases or remains constant, a case which has already been sufficiently discussed.

§ **142. Case II. Isothermal Process.**—The temperature T being kept constant, (70) passes into

$$d(U - T\Phi) \leqq W \quad . \quad . \quad . \quad (70a)$$

i.e. the increment of the quantity $(U - T\Phi)$ is smaller than, or, in the limit, equal to, the work done on the system. Since in thermochemistry the final state is always reduced to the temperature of the initial state, this theorem is well adapted for application to chemical processes.

Putting $$U - T\Phi = F, \quad . \quad . \quad . \quad (71)$$

we have, for reversible isothermal changes :

$$dF = W$$

and, on integrating,

$$F_2 - F_1 = \Sigma W \quad . \quad . \quad . \quad (72)$$

For finite reversible isothermal changes the total work done on the system is equal to the increase of F; or, the total work performed by the system is equal to the decrease of F, and, therefore, depends only on the initial and final states of the system. When $F_1 = F_2$, as in cyclic processes, the external work is zero (cf. § 139).

The function F, thus bearing the same relation to the external work that the energy U does to the sum of the external heat and work, has been called by H. v. Helmholtz the *free energy* (freie Energie) of the system. (It should rather be called " free energy for isothermal processes.") Corresponding to this, he calls U the *total energy* (Gesammtenergie), and the difference

$$U - F = T\Phi = G$$

the *latent energy* (gebundene Energie) of the system. The change of the latter in reversible isothermal processes gives the heat absorbed. By reversible isothermal processes the principle of the conservation of energy

$$U_2 - U_1 = \Sigma W + \Sigma Q$$

breaks up into two parts, the law of the free energy :

$$F_2 - F_1 = \Sigma W,$$

and the law of the latent energy :

$$G_2 - G_1 = \Sigma Q.$$

It should be observed, however, that this division is applicable only to isothermal changes.

In irreversible processes, by (70a), $dF < W$, and on integrating we have

$$F_2 - F_1 < \Sigma W \quad \cdots \quad \cdots \quad (73)$$

The free energy increases by a less amount than what corresponds to the work done on the system. The results for reversible and irreversible processes may be stated thus. In irreversible isothermal processes the work done on the system is more, or the work done by the system is less, than it would be if the same change were brought about by a reversible process, for in that case it would be the difference of the free energies at the beginning and end of the process (72).

Hence, any reversible transformation of the system from one state to another yields the maximum amount of work that can be gained by any isothermal process between those two states. In all irreversible processes a certain amount of work is lost, viz. the difference between the maximum work to be gained (the decrease of the free energy) and the work actually gained.

The fact that, in the above, irreversible as well as reversible processes between the same initial and final states were considered, does not contradict the proposition that between two states of a system either only reversible or only irreversible processes are possible, if no external changes are to remain in other bodies. In fact, the process here discussed involves such changes in the surrounding medium ; for, in order to keep the system at constant temperature, an exchange of heat between it and the surrounding medium must take place in one direction or the other.

§ **143.** If the work done during an isothermal process vanish, as is practically the case in most chemical reactions, we have

$$\Sigma W = 0,$$

and, by (73), $\qquad F_2 - F_1 < 0,$

i.e. the free energy decreases. The amount of this decrease may be used as a measure of the work done by the forces (chemical affinity) causing the process, for the same is not available for external work.

For instance, let an aqueous solution of some non-volatile salt be diluted isothermally, the heat of dilution being furnished or received by a heat-reservoir according as the energy, U_2, of the diluted solution (final state) is greater or less than the sum, U_1, of the energies of the undiluted solution and the water added (initial state). The free energy, F_2, of the diluted solution, on the other hand, is necessarily smaller than the sum, F_1, of the free energies of the undiluted solution and the water added. The amount of the decrease of the free energy, or the work done by the " affinity of the solution for water " during the process of dilution may be measured. For this purpose, the dilution should be performed in some reversible isothermal manner, when, according to (72), the quantity to be measured is actually gained in the form of external work. For instance, evaporate the water, which is to be added, infinitely slowly under the pressure of its saturated vapour. When it has all been changed to water vapour, allow the latter to expand isothermally and reversibly until its density equals that which saturated water vapour would possess at that temperature when in contact with the solution. Now establish lasting contact between the water vapour and the solution, whereby the equilibrium will not be disturbed. Finally, by isothermal compression, condense the water vapour infinitely slowly when in direct contact with the solution. It will then be uniformly distributed throughout the latter. Such a process, as here described, is composed only of states of equilibrium. Hence it is reversible, and the external work thereby gained represents at the same time the decrease of the free energy, $F_2 - F_1$, which takes place on directly mixing the solution and the water. Each method of bringing the water, isothermally and reversibly, into the solution must, of course, give the same value for this difference. In this way Helmholtz calculated the electromotive force of a concentration cell from the vapour pressure of the solution.

As a further example, we shall take a mixture of hydrogen and oxygen which has been exploded by means of an electric spark. The spark acts only the secondary part of a release, its energy being negligible in comparison with the energies obtained by the reaction. The work of the chemical affinities in this process is equal to the mechanical work that might be gained by chemically combining the oxygen and hydrogen in some reversible and isothermal way. Dividing this quantity by the number of oxidized molecules of hydrogen, we obtain a measure of the force with which a molecule of hydrogen tends to become oxidized. This definition of chemical force, however, has only a meaning in so far as it is connected with that work.

§ **144.** In chemical processes the changes of the first term, U, of the expression for the free energy (71), frequently far surpass those of the second, $T\Phi$. Under such circumstances, instead of the decrease of F, that of U, *i.e.* the heat of reaction, may be considered as a measure of the chemical work. This leads to the proposition that chemical reactions, in which there is no external work, take place in such a manner as to give the greatest evolution of heat (Berthelot's principle). For high temperatures, where T is large, and for gases and dilute solutions, where Φ is large, the term $T\Phi$ can no longer be neglected without considerable error. In these cases, therefore, chemical changes often do take place in such a way as to increase the total energy, *i.e.* with the absorption of heat.

§ 145. It should be borne in mind that all these propositions refer only to isothermal processes. To answer the question as to how the free energy acts in other processes, it is only necessary to form the differential of (71) viz. :

$$dF = dU - Td\Phi - \Phi dT$$

and to substitute in the general relation (70). We have then

$$dF \leqq W - \Phi dT$$

for any physical or chemical process. This shows that, with change of temperature, the relation between the external work and the free energy is far more complicated. This relation cannot, in general, be used with advantage.

§ 146. We shall now compute the value of the free energy of a perfect gas. Here, according to (35),

$$U = Mu = M(c_v T + b) \ (b \text{ constant}),$$

and, by (52),

$$\Phi = M\phi = M(c_v \log T + \frac{R}{m} \log v + a) \ (a \text{ constant}).$$

Substituting in (71), we obtain

$$F = M\{T(c_v - a - c_v \log T) - \frac{RT}{m} \log v + b\} \quad (74)$$

which contains an arbitrary linear function of T.

The constant b cannot be determined from thermodynamics; on the other hand, a method of calculating the constant a is given by Nernst's heat theorem (§ 287).

For isothermal changes of the gas, we have, by § 142,

$$dF \leqq W,$$

or, by (74), since T is const.,

$$dF = - \frac{MTR}{m} \cdot \frac{dv}{v} = - pdV \leqq W.$$

If the change be reversible, the external work on the gas is $W = - pdV$, but if it be irreversible, then the sign of inequality shows that the work of compression is greater, or that of expansion smaller, than in a reversible process.

§ 147. Case III. Isothermal–isobaric Process.—

If, besides the temperature T, the external pressure p be also kept constant, then the external work is given by the formula,

$$W = - pdV,$$

and the left-hand side of (69) becomes a complete differential :

$$d\left(\Phi - \frac{U + pV}{T}\right) \geqq 0.$$

In this case, it may be stated that for finite changes the function,

$$\Phi - \frac{U + pV}{T} = \Psi, \quad \ldots \quad (75)$$

must increase, and will remain constant only in the limit when the change is reversible.*

§ 148. Conditions of Equilibrium.—

The most general condition of equilibrium for any system of bodies is derived from the proposition that no change can take place in the system if it be impossible to satisfy the condition necessary for a change.

Now, by (69), for any actual change of the system,

$$d\Phi - \frac{dU - W}{T} > 0.$$

The sign of equality is omitted, because it refers to ideal changes which do not actually occur in nature. Equilibrium is, therefore, maintained if the fixed conditions imposed on the system be such that they will permit only changes in which

$$\delta\Phi - \frac{\delta U - W}{T} \leq 0.$$

---

* Multiplying (75) by $-$ T, we get the *thermodynamic potential at constant pressure,*

$$U + pV - T\phi,$$

for which, so long as T remains constant, the same propositions hold as for the function Ψ. However, the important equation 153 in § 211, which shows how the equilibrium depends on temperature, pressure and the masses of the independent constituents of the system, can be more conveniently deduced from the function Ψ than from the thermodynamic potential.

Here δ is used to signify a virtual infinitely small change, in contrast to d, which corresponds to an actual change.

§ **149.** In most of the cases subsequently discussed, if any given virtual change be compatible with the fixed conditions of the system, its exact opposite is also, and is represented by changing the sign of all variations involved. This is true if the fixed conditions be expressed by equations, not by inequalities. Assuming this to be the case, if we should have, for any particular virtual change,

$$\delta\Phi - \frac{\delta U - W}{T} < 0,$$

which, by (69), would make its occurrence in nature impossible, its opposite would conform to the condition for actual changes (69), and could therefore take place in nature. To ensure equilibrium in such cases, it is necessary, therefore, that, for any virtual change compatible with the fixed conditions,

$$\delta\Phi - \frac{\delta U - W}{T} = 0 \quad . \quad . \quad . \quad . \quad (76)$$

This equation contains a condition always sufficient, but, as we have seen, not always necessary to its full extent, for the maintenance of equilibrium. As a matter of experience, equilibrium will occasionally subsist when equation (76) is not fulfilled, even though the fixed conditions permit of a change of sign of all variations. This is to say, that occasionally a certain change will not take place in nature, though it satisfy the fixed conditions as well as the demands of the second law. Such cases lead to the conclusion that in some way the setting in of a change meets with a certain resistance, which, on account of the direction in which it acts, has been termed inertia resistance, or passive resistance. States of equilibrium of this description are always unstable. Often a very small disturbance, not comparable in size with the quantities within the system, suffices to produce the change, which under these conditions often occurs with great violence. We have examples of this in overcooled liquids, supersaturated vapour, supersaturated solutions, explosive substances, etc. We shall henceforth discuss mainly the conditions of stable equilibrium deducible from (76).

This equation may, under certain circumstances, be expressed in the form of a condition for a maximum or minimum. This can be done when, and only when, the conditions imposed upon the system are such that the left-hand side of (76) represents the variation of a definite finite function. The most important of these cases are dealt with separately in the following paragraphs. They correspond exactly to the propositions which we have already deduced for special cases. From these propositions it may at once be seen whether it is a case of a maximum or a minimum.

§ **150. First Case** (§ 141).—If no exchange of heat take place with the surrounding medium, the first law gives

$$\delta U = W,$$

hence, by (76),
$$\delta \Phi = 0 \quad . \quad . \quad . \quad . \quad . \quad (77)$$

Among all the states of the system which can proceed from one another by adiabatic processes, the state of equilibrium is distinguished by a maximum of the entropy. Should there be several states in which the entropy has a maximum value, each one of them is a state of equilibrium; but if the entropy be greater in one than in all the others, then that state represents absolutely stable equilibrium, for it could no longer be the starting-point of any change whatsoever.

§ **151. Second Case** (§ 142).—If the temperature be kept constant, equation (76) passes into

$$\delta\left(\Phi - \frac{U}{T}\right) + \frac{W}{T} = 0,$$

and, by (71),
$$- \delta F = - W.$$

Among all the states which the system may assume at a given temperature, a state of equilibrium is characterized by the fact that the free energy of the system cannot decrease without performing an equivalent amount of work.

If the external work be a negligible quantity, as it is when the volume is kept constant or in numerous chemical processes, then W = 0, and the condition of equilibrium becomes

$$\delta F = 0,$$

i.e. among the states which can proceed from one another by isothermal processes, without the performance of external work, the state of most stable equilibrium is distinguished by an absolute minimum of the free energy.

§ **152. Third Case** (§ 147).—Keeping the temperature T and the pressure p constant and uniform, we have

$$W = -p\delta V, \quad . \quad . \quad . \quad . \quad (78)$$

and the condition of equilibrium (76) becomes

$$\delta\left(\Phi - \frac{U + pV}{T}\right) = 0,$$

or, by (75), $\qquad \delta\Psi = 0 \quad . \quad . \quad . \quad . \quad (79)$

i.e. at constant temperature and constant pressure, the state of most stable equilibrium is characterized by an absolute maximum of the function Ψ.

We shall in the next part consider states of equilibrium of various systems by means of the theorems we have just deduced, going from simpler to more complicated cases.

152A. In the mathematical treatment of thermodynamical problems of equilibrium the choice of the independent variable is of first importance. When this choice has been made, characteristic functions come into prominence. Herein lies the difference in the form of the presentation by different authors. In each case there exists, as Massieu pointed out, a characteristic function, from which, by simple differentiation, all thermodynamical properties of the system can be determined uniquely. In fact, on keeping certain independent variables constant, the function is always that one whose maximum or minimum characterizes the thermodynamical equilibrium according to the laws of the preceding paragraphs.

If, for example, the energy U and the volume V are the

independent variables, then the entropy is the characteristic function. We find directly from the equation :

$$d\Phi = \frac{d\mathrm{U} + p d\mathrm{V}}{\mathrm{T}} = \frac{\partial\Phi}{\partial\mathrm{U}}d\mathrm{U} + \frac{\partial\Phi}{\partial\mathrm{V}}d\mathrm{V},$$

that
$$\left(\frac{\partial\Phi}{\partial\mathrm{U}}\right)_\mathrm{V} = \frac{1}{\mathrm{T}} \text{ and } \left(\frac{\partial\Phi}{\partial\mathrm{V}}\right)_\mathrm{U} = \frac{p}{\mathrm{T}}.$$

If Φ is a known function of U and V, then

$$\mathrm{T} = \frac{1}{\dfrac{\partial\Phi}{\partial\mathrm{U}}}, \quad p = \frac{\dfrac{\partial\Phi}{\partial\mathrm{V}}}{\dfrac{\partial\Phi}{\partial\mathrm{U}}},$$

and hence Φ and U can be found as functions of T and p.

If, on the other hand, V and T are chosen as independent variables, then the free energy, F, is the characteristic function. From (71), it follows that

$$d\mathrm{F} = d\mathrm{U} - \mathrm{T}d\Phi - \Phi d\mathrm{T}$$

or
$$d\mathrm{F} = - p d\mathrm{V} - \Phi d\mathrm{T},$$

hence
$$\left(\frac{\partial\mathrm{F}}{\partial\mathrm{V}}\right)_\mathrm{T} = - p \text{ and } \left(\frac{\partial\mathrm{F}}{\partial\mathrm{T}}\right)_\mathrm{V} = - \Phi.$$

If F is a known function of V and T, then we have the following expressions :

$$p = - \frac{\partial\mathrm{F}}{\partial\mathrm{V}}, \ \Phi = - \frac{\partial\mathrm{F}}{\partial\mathrm{T}}, \ \mathrm{U} = \mathrm{F} + \mathrm{T}\Phi = \mathrm{F} - \mathrm{T}\frac{\partial\mathrm{F}}{\partial\mathrm{T}}. \quad (79a)$$

If, lastly, p and T are chosen as independent variables, then the characteristic function is

$$\Psi = \Phi - \frac{\mathrm{U} + p\mathrm{V}}{\mathrm{T}}.$$

It follows that

$$d\Psi = d\Phi - \frac{d\mathrm{U} + p d\mathrm{V} + \mathrm{V}dp}{\mathrm{T}} + \frac{\mathrm{U} + p\mathrm{V}}{\mathrm{T}^2}d\mathrm{T},$$

or
$$d\Psi = - \frac{\mathrm{V}}{\mathrm{T}}dp + \frac{\mathrm{U} + p\mathrm{V}}{\mathrm{T}^2}d\mathrm{T},$$

hence $\quad \left(\dfrac{\partial \Psi}{\partial p}\right)_T = -\dfrac{V}{T}$ and $\left(\dfrac{\partial \psi}{\partial T}\right)_p = \dfrac{U + pV}{T^2}.$. (79b)

If Ψ is a known function of p and T, then we have

$$V = -T\dfrac{\partial \psi}{\partial p}, \ U = T\Big(T\dfrac{\partial \psi}{\partial T} + p\dfrac{\partial \psi}{\partial p}\Big), \quad . \quad (79c)$$

and $\quad \Phi = \psi + T\dfrac{\partial \psi}{\partial T}.$

From these, all thermodynamical properties of the system in question are determined uniquely.

§ 152B. Although the derivation of all thermodynamical properties from the characteristic functions Φ, F, and ψ is so important in principle, yet in practice these functions are of direct use only when it is really possible to find the expression for the particular characteristic function in terms of the independent variables. It is of particular importance then to find out how far this expression can be obtained by direct heat measurements.

We shall take T and p as independent variables, since heats of reaction are usually determined under the external conditions of temperature and pressure constant. The change in Gibbs' heat function (§ 100)

$$H = U + pV$$

gives the heat of reaction of a physical chemical process. H is uniquely determined from the characteristic function ψ.

By equation (79b),

$$H = T^2\dfrac{\partial \psi}{\partial T} \quad . \quad . \quad . \quad . \quad (79d)$$

If, on the other hand, ψ is to be determined from H, after H has been experimentally measured, then the last equation has to be integrated at constant pressure :

$$\psi = \int\dfrac{H}{T}dT \ . \quad . \quad . \quad . \quad (79e)$$

In this integral two additive terms remain undetermined :

first, the constant of integration; second, a term which depends on the fact that in the expression for H, as in that for U, an additive constant is undetermined.   The first term is of the form a, the second of the form $\dfrac{b}{T}$.   Accordingly there remains in the expression for the entropy Φ (79c) an additive constant a, and in the expression for the free energy $F = U - T\Phi$ an additive term of the form $b - aT$ undetermined.   This we have already seen in the special case of a perfect gas [§ 146, equation (74)].   In general, a and b are independent of the pressure.   For further information about the determination of ψ from heat measurements see § 210.

§ **152c.**   If we differentiate the equation (79d) with respect to p keeping the temperature constant, we obtain with the help of (79b)

$$\frac{\partial H}{\partial p} = -\ T^2 \frac{\partial}{\partial T}\!\left(\frac{V}{T}\right) = V - T\frac{\partial V}{\partial T}\ . \quad . \quad (79f)$$

This equation gives a quite general relation between heat of reaction and thermal changes of volume, which could be used as an experimental test of the second law.   In any isothermal isobaric reaction the heat added is, by § 105,

$$Q = H_2 - H_1,$$

where 1 and 2 denote the initial and final states of the system. If, further, we call the change of volume that takes place on the reaction :

$$V_2 - V_1 = \Delta V,$$

then it follows, from (79f), that

$$\frac{\partial Q}{\partial p} = -\ T^2 \frac{\partial}{\partial T}\!\left(\frac{\Delta V}{T}\right). \quad . \quad . \quad (79g)$$

It follows from this, for example, that, if the heat of reaction of an isothermal reaction is independent of the pressure, the corresponding change of volume is proportional to the absolute temperature $\left(\dfrac{\Delta V}{T} = \text{const.}\right)$ and *vice versâ*.

PART IV.

CHAPTER I.

HOMOGENEOUS SYSTEMS.

§ 153. LET the state of a homogeneous system be determined, as hitherto, by its mass, M; its temperature, T; and either its pressure, p, or its specific volume, $v = \dfrac{V}{M}$. For the present, besides M, let T and v be the independent variables. Then the pressure p, the specific energy $u = \dfrac{U}{M}$, and the specific entropy $\phi = \dfrac{\Phi}{M}$ are functions of T and v, the definition of the specific entropy (61) being

$$d\phi = \frac{du + pdv}{T} = \frac{1}{T}\left(\frac{\partial u}{\partial T}\right)_v dT + \frac{\left(\frac{\partial u}{\partial v}\right)_T + p}{T}dv.$$

On the other hand,

$$d\phi = \left(\frac{\partial \phi}{\partial T}\right)_v dT + \left(\frac{\partial \phi}{\partial v}\right)_T dv.$$

Therefore, since dT and dv are independent of each other,

$$\left(\frac{\partial \phi}{\partial T}\right)_v = \frac{1}{T}\left(\frac{\partial u}{\partial T}\right)_v. \quad \cdot \quad \cdot \quad \cdot \quad \cdot \quad (79h)$$

and

$$\left(\frac{\partial \phi}{\partial v}\right)_T = \frac{\left(\frac{\partial u}{\partial v}\right)_T + p}{T}.$$

These two equations lead to an experimental test of the

second law; for, differentiating the first with respect to v, the second with respect to T, we have

$$\frac{\partial^2 \phi}{\partial T \partial v} = \frac{1}{T} \cdot \frac{\partial^2 u}{\partial T \partial v} = \frac{\dfrac{\partial^2 u}{\partial T \partial v} + \left(\dfrac{\partial p}{\partial T}\right)_v}{T} - \frac{\left(\dfrac{\partial u}{\partial v}\right)_T + p}{T^2}$$

or

$$\left(\frac{\partial u}{\partial v}\right)_T = T\left(\frac{\partial p}{\partial T}\right)_v - p . \quad . \quad . \quad . \quad (80)$$

By this and equation (24), the above expressions for the differential coefficients of ϕ become :

$$\left(\frac{\partial \phi}{\partial T}\right)_v = \frac{c_v}{T}; \text{ and } \left(\frac{\partial \phi}{\partial v}\right)_T = \left(\frac{\partial p}{\partial T}\right)_v \quad . \quad . \quad (81)$$

If we differentiate the first of these equations with respect to v and the second with respect to T and equate, we get

$$\left(\frac{\partial c_v}{\partial v}\right)_T = T\left(\frac{\partial^2 p}{\partial T^2}\right)_v \quad . \quad . \quad . \quad (81a)$$

This equation shows how the variation of the specific heat at constant volume with change of volume is related to the change of the thermal pressure coefficient with change of temperature. Both are remarkably small in the case of perfect gases.

§ 154. Equation (80), together with (28) of the first law, gives the relation :

$$c_p - c_v = T\left(\frac{\partial p}{\partial T}\right)_v \cdot \left(\frac{\partial v}{\partial T}\right)_p \quad . \quad . \quad (82)$$

which is useful either as a test of the second law or for the calculation of c_v when c_p is given. But since in many cases $\left(\frac{\partial p}{\partial T}\right)_v$ cannot be directly measured, it is better to introduce the relation (6), and then

$$c_p - c_v = - T\left(\frac{\partial p}{\partial v}\right)_T \cdot \left(\frac{\partial v}{\partial T}\right)_p^2 \quad . \quad . \quad (83)$$

As $\left(\frac{\partial p}{\partial v}\right)_T$ is necessarily negative, c_p is always greater than c_v, except in the limiting case, when the coefficient of expansion is $= 0$, as in the case of water at $4°$ C.; then $c_p - c_v = 0$.

As an example, we may calculate the specific heat at constant volume, c_v, of mercury at $0°$ C. from the following data :

$$c_p = 0\cdot0333; \; T = 273°;$$

$$\left(\frac{\partial p}{\partial v}\right)_T = -\frac{1013250}{0\cdot0000039\,.\,v},$$

where the denominator is the coefficient of compressibility in atmospheres (§ 15); the numerator, the pressure of an atmosphere in absolute units (§ 7); $v = \dfrac{1}{13\cdot596}$, the volume of 1 gr. of mercury at $0°$ C. ; $\left(\dfrac{\partial v}{\partial T}\right)_p = 0\cdot0001812\,.\,v$, the coefficient of thermal expansion (§ 15).

To obtain c_v in calories, it is necessary to divide by the mechanical equivalent of heat, $4\cdot19 \times 10^7$ (§ 61). Thus we obtain from (83)

$$c_p - c_v = \frac{273 \times 1013250 \times 0\cdot0001812^2}{0\cdot0000039 \times 13\cdot596 \times 4\cdot19 \times 10^7} = 0\cdot0041,$$

whence, from the above value for c_p,

$$c_v = 0\cdot0292, \text{ and } \frac{c_p}{c_v} = 1\cdot1.$$

§ 155. This method of calculating the difference of the specific heats $c_p - c_v$, applicable to any substance, discloses at the same time the order of magnitude of the different influences to which this quantity is subject. According to equation (28) of the first law, the difference of the specific heats is

$$c_p - c_v = \left\{\left(\frac{\partial u}{\partial v}\right)_T + p\right\}\left(\frac{\partial v}{\partial T}\right)_p$$

The two terms of this expression, $\left(\dfrac{\partial u}{\partial v}\right)_T\left(\dfrac{\partial v}{\partial T}\right)_p$ and $p\left(\dfrac{\partial v}{\partial T}\right)_p$, depend on the rate of change of the energy with the volume, and on the external work performed by the expansion respectively. In order to find which of these two terms has the greater influence on the quantity $c_p - c_v$, we shall find the ratio of the first to the second :

$$\frac{1}{p} \cdot \left(\frac{\partial u}{\partial v}\right)_{\mathrm{T}},$$

or, by (80), $$\frac{\mathrm{T}}{p} \cdot \left(\frac{\partial p}{\partial \mathrm{T}}\right)_v - 1 \quad . \quad . \quad . \quad . \quad . \quad (84)$$

or, by (6), $$-\frac{\mathrm{T}}{p}\left(\frac{\partial v}{\partial \mathrm{T}}\right)_p \left(\frac{\partial p}{\partial v}\right)_{\mathrm{T}} - 1.$$

A glance at the tables of the coefficients of thermal expansion and of the compressibility of solids and liquids shows that, in general, the first term of this expression is a large number, making the second, -1, a negligible quantity. For mercury at $0°$, *e.g.*, the above data give the first term to be

$$273 \times \frac{0 \cdot 0001812}{0 \cdot 0000039} = 12700.$$

Water at $4°$ C. is an exception.

It follows that, for solids and liquids, the difference $c_p - c_v$ depends rather on the relation between the energy and the volume than on the external work of expansion. For perfect gases the reverse is the case, since the internal energy is independent of the volume, *i.e.*—

$$\left(\frac{\partial u}{\partial v}\right)_{\mathrm{T}} = 0.$$

During expansion, therefore, the influence of the internal energy vanishes in comparison with that of the external work; in fact, the expression (84) vanishes for the characteristic equation of a perfect gas. With ordinary gases, however, both the internal energy and the external work must be considered.

§ 156. The sum of both these influences, *i.e.* the whole expression, $c_p - c_v$, may be said to have a small value for most solids and liquids; thus the ratio $\frac{c_p}{c_v} = \gamma$ is but slightly greater than unity. This means that in solids and liquids the energy depends far more on the temperature than on the volume. For gases, γ is large; and, in fact, the fewer the number of atoms in a molecule of the gas, the larger does it become. Hydrogen, oxygen, and most gases with diatomic molecules have $\gamma = 1\cdot41$ (§ 87). The largest observed value of γ is that found by Kundt and Warburg for the monatomic vapour of mercury, viz. $1\cdot667$, which is also the value for the monatomic permanent gases.

§ 157. For many applications of the second law it is convenient to introduce p instead of v as an independent variable. We have, by (61),

$$d\phi = \frac{du + pdv}{T} = \left\{ \left(\frac{\partial u}{\partial T}\right)_p + p\left(\frac{\partial v}{\partial T}\right)_p \right\} \frac{dT}{T} + \left\{ \left(\frac{\partial u}{\partial p}\right)_T + p\left(\frac{\partial v}{\partial p}\right)_T \right\} \frac{dp}{T}.$$

On the other hand,

$$d\phi = \left(\frac{\partial \phi}{\partial T}\right)_p dT + \left(\frac{\partial \phi}{\partial p}\right)_T dp,$$

whence, $\qquad \left(\frac{\partial \phi}{\partial T}\right)_p = \dfrac{\left(\dfrac{\partial u}{\partial T}\right)_p + p\left(\dfrac{\partial v}{\partial T}\right)_p}{T}$

and $\qquad \left(\frac{\partial \phi}{\partial p}\right)_T = \dfrac{\left(\dfrac{\partial u}{\partial p}\right)_T + p\left(\dfrac{\partial v}{\partial p}\right)_T}{T}$

Differentiating the first of these with respect to p, the second with respect to T, we get,

$$\frac{\partial^2 \phi}{\partial T \partial p} = \frac{\dfrac{\partial^2 u}{\partial T \partial p} + p\dfrac{\partial^2 v}{\partial T \partial p} + \left(\dfrac{\partial v}{\partial T}\right)_p}{T} = \frac{\dfrac{\partial^2 u}{\partial T \partial p} + p\dfrac{\partial^2 v}{\partial T \partial p}}{T} - \frac{\left(\dfrac{\partial u}{\partial p}\right)_T + p\left(\dfrac{\partial v}{\partial p}\right)_T}{T^2}$$

whence $\qquad \left(\frac{\partial u}{\partial p}\right)_T = - T\left(\frac{\partial v}{\partial T}\right)_p - p\left(\frac{\partial v}{\partial p}\right)_T.$

The differential coefficients of ϕ become, then, by (26),

$$\left(\frac{\partial\phi}{\partial T}\right)_p = \frac{c_p}{T}; \quad \cdots \quad (84a)$$

and

$$\left(\frac{\partial\phi}{\partial p}\right)_T = -\left(\frac{\partial v}{\partial T}\right)_p. \quad \cdots \quad (84b)$$

Finally, differentiating the first of these with respect to p, the second with respect to T, and equating, we have

$$\left(\frac{\partial c_p}{\partial p}\right)_T = - T\left(\frac{\partial^2 v}{\partial T^2}\right)_p. \quad \cdots \quad (85)$$

This equation, which could be obtained directly by differentiating the general equation ($79f$) in § 152c with respect to T, contains only quantities which can be directly measured. It establishes a relation between the rate of change of the specific heat at constant pressure with pressure and the rate of change of the coefficient of thermal expansion with temperature. This relation depends upon the deviations from Gay-Lussac's law. Compare this equation (85) with ($81a$).

§ **158.** By means of the relations furnished by the second law we may also draw a further conclusion from Thomson and Joule's experiments (§ 70), in which a gas was slowly pressed through a tube plugged with cotton wool. The interpretation in § 70 was confined to their bearing on the properties of perfect gases. It has been mentioned that the characteristic feature of these experiments consists in giving to a gas, without adding or withdrawing heat, ($Q = 0$) an increase of volume, $V_2 - V_1$, or $v_2 - v_1$ per unit mass, while the external work done per unit mass is represented by

$$p_1 v_1 - p_2 v_2 = W.$$

This expression vanishes in the case of perfect gases, since then the temperature remains constant. In the case of actual gases, it follows, in general, from equation (17) of the first law that

$$u_2 - u_1 = p_1 v_1 - p_2 v_2,$$

or :

$$u_2 + p_2 v_2 = u_1 + p_1 v_1.$$

If now we introduce the heat function per unit mass $\dfrac{H}{M} = h$, we have :

$$h_2 = h_1.$$

In the Joule-Thomson experiment the energy u does not in general remain constant, but the heat function $h = u + pv$ remains constant. If for the sake of simplicity we take the pressures on the two sides of the plug to be very slightly different, and denote the difference of all quantities on the two sides of the plug by Δ, then the last equation may be written :

$$\Delta h = 0 = \Delta u + p\Delta v + v\Delta p$$

or from (61) :

$$T\Delta\phi + v\Delta p = 0,$$

hence :

$$T\left(\frac{\partial\phi}{\partial T}\right)_p \Delta T + \left\{T\left(\frac{\partial\phi}{\partial p}\right)_T + v\right\}\Delta p = 0.$$

This gives, with the help of (84a) and (84b),

$$c_p\Delta T + \left\{v - T\left(\frac{\partial v}{\partial T}\right)_p\right\}\Delta p = 0$$

and

$$\Delta T = \frac{T\left(\dfrac{\partial v}{\partial T}\right)_p - v}{c_p}\Delta p \quad . \quad . \quad . \quad (86)$$

By means of this simple equation, the change of temperature (ΔT) of the gas in Thomson and Joule's experiments, for a difference of pressure Δp, may be found from its specific heat at constant pressure, c_p, and its deviation from Gay-Lussac's law. If, under constant pressure, v were proportional to T, as in Gay-Lussac's law, then, by equation (86), $\Delta T = 0$ as is really the case for perfect gases.

§ **158A.** If we take van der Waals' equation as the characteristic equation, then equation (86) gives

$$\Delta T = \frac{2a(v - b)^2 - RTbv^2}{RTv^3 - 2a(v - b)^2} \cdot \frac{v\Delta p}{c_p}.$$

If a and b become sufficiently small,

$$\Delta T = \left(\frac{2a}{RT} - b\right)\frac{\Delta p}{c_p},$$

a relation which agrees very closely with experiment. For most gases the expression within the bracket is, at ordinary, not too high temperatures, positive. This means that the effect is a cooling effect, since Δp is always negative. C. von Linde based the construction of his liquid air machine on this fact.

In the case of hydrogen the bracket is negative at mean temperatures on account of the smallness of a. The positive term prevails only if the temperature is lowered below $-80°$ C. The so-called *point of inversion* in which the effect changes sign, lies according to the last equation at

$$T = \frac{2a}{Rb}$$

Since this equation is only approximately true, the position of the point of inversion depends, in general, on the pressure.*

§ 159. Thomson and Joule embraced the results of their observations in the formula

$$\Delta T = \frac{\alpha}{T^2}\Delta p,$$

where α is a constant. If we express p in atmospheres, we have, for air,

$$\alpha = 0.276 \times (273)^2.$$

No doubt the formula is only approximate. Within the region of its validity we get, from 86,

$$T\left(\frac{\partial v}{\partial T}\right)_p - v = c_p\frac{\alpha}{T^2} \quad . \quad . \quad . \quad . \quad (87)$$

and, differentiating with respect to T,

$$T\left(\frac{\partial^2 v}{\partial T^2}\right)_p = \frac{\alpha}{T^2}\left(\frac{\partial c_p}{\partial T}\right)_p - \frac{2\alpha c_p}{T^3}$$

* F. A. Schulze, *Ann. d. Phys*, **49**, p. 585, 1916.

whence, by the relation (85),

$$\left(\frac{\partial c_p}{\partial p}\right)_{\mathrm{T}} + \frac{\alpha}{\mathrm{T}^2}\left(\frac{\partial c_p}{\partial \mathrm{T}}\right)_p - \frac{2\alpha c_p}{\mathrm{T}^3} = 0.$$

The general solution of this differential equation is

$$c_p = \mathrm{T}^2 . f(\mathrm{T}^3 - 3\alpha p),$$

where f denotes an arbitrary function of its argument, $\mathrm{T}^3 - 3\alpha p$.

If we now assume that, with diminishing values of p, the gas, at any temperature, approaches indefinitely near the ideal state, then, when $p = 0$, c_p becomes a constant $= c_p^{(0)}$ (for air, $c_p^{(0)} = 0.238$ calorie). Hence, generally,

$$c_p = c_p^{(0)}\mathrm{T}^2(\mathrm{T}^3 - 3\alpha p)^{-\frac{2}{3}}$$

or

$$c_p = \frac{c_p^{(0)}}{\left(1 - \dfrac{3\alpha p}{\mathrm{T}^3}\right)^{\frac{2}{3}}} \quad \quad . \quad . \quad . \quad . \quad (88)$$

This expression for c_p will serve further to determine v in terms of T and p. It follows from (87) that

$$\mathrm{T}^2\frac{\partial}{\partial \mathrm{T}}\left(\frac{v}{\mathrm{T}}\right)_p = \frac{\alpha c_p}{\mathrm{T}^2} = \frac{\alpha c_p^{(0)}}{(\mathrm{T}^3 - 3\alpha p)^{\frac{2}{3}}},$$

whence

$$\frac{v}{\mathrm{T}} = \alpha c_p^{(0)}\int \frac{d\mathrm{T}}{\mathrm{T}^4\left(1 - \dfrac{3\alpha p}{\mathrm{T}^3}\right)^{\frac{2}{3}}},$$

or

$$v = \frac{c_p^{(0)}\mathrm{T}}{3p}\left(\sqrt[3]{1 - \frac{3\alpha p}{\mathrm{T}^3}} + \beta\right). \quad . \quad (89)$$

This is the characteristic equation of the gas. The constant of integration, β, may be determined from the density of the gas at $0°$ C. under atmospheric pressure. Equations (88) and (89), like Thomson and Joule's formula, are valid only within certain limits. It is, however, of theoretical interest to see how the different relations necessarily follow from one another.*

* These considerations have been worked out and applied to experimental data by R. Plank, *Physikalische Zeitschrift*, 11, p. 633, 1910, and 17, p. 521, 1916.
See also L. Schames, *Physikalische Zeitschrift*, 18, p. 30. 1917.

§ 160. A further, theoretically important application of the second law is the determination of the absolute temperature of a substance by a method independent of the deviations of actual gases from the ideal state. In § 4 we defined temperature by means of the gas thermometer, but had to confine that definition to the cases in which the readings of the different gas thermometers (hydrogen, air, etc.) agree within the limits of the desired accuracy. For all other cases (including mean temperatures, when a high degree of accuracy is desired) we postponed the definition of absolute temperature. Equation (80) enables us to give an exact definition of absolute temperature, entirely independent of the behaviour of special substances.

Given the temperature readings, t, of any arbitrary thermometer (mercury-thermometer, or the scale deflection of a thermo-element, or of a bolometer), our problem is to reduce the thermometer to an absolute one, or to express the absolute temperature T as a function of t. We may by direct measurement find how the behaviour of some appropriate substance, *e.g.* a gas, depends on t and on either v or p. Introducing, then, t and v as the independent variables in (80) instead of T and v, we obtain

$$\left(\frac{\partial u}{\partial v}\right)_t = \mathrm{T}\left(\frac{\partial p}{\partial t}\right)_v \cdot \frac{dt}{d\mathrm{T}} - p,$$

where $\left(\frac{\partial u}{\partial v}\right)_t$, p, and $\left(\frac{\partial p}{\partial t}\right)_v$ represent functions of t and v, which can be experimentally determined. The equation can then be integrated thus :

$$\int \frac{\partial \mathrm{T}}{\mathrm{T}} = \int \frac{\left(\frac{\partial p}{\partial t}\right)_v dt}{\left(\frac{\partial u}{\partial v}\right)_t + p}.$$

If we further stipulate that at the freezing-point of water, where $t = t_0$, $\mathrm{T} = \mathrm{T}_0 = 273$, then,

$$\log \frac{T}{T_0} = \int_{t_0}^{t} \frac{\left(\frac{\partial p}{\partial t}\right)_v dt}{\left(\frac{\partial u}{\partial v}\right)_t + p}.$$

This completely determines T as a function of t. It is evident that the volume, v, no longer enters into the expression under the sign of integration. This conclusion from the theory can at the same time be used to test the correctness of the second law of thermodynamics.

§ 161. The numerator of this expression may be found directly from the characteristic equation of the substance. The denominator, however, depends on the amount of heat which the substance absorbs during isothermal reversible expansion. For, by (22) of the first law, the ratio of the heat absorbed during isothermal reversible expansion to the change of volume is

$$\left(\frac{q}{dv}\right)_t = \left(\frac{\partial u}{\partial v}\right)_t + p.$$

§ 162. Instead of measuring the quantity of heat absorbed during isothermal expansion, it may be more convenient, for the determination of the absolute temperature, to experiment on the changes of temperature of a slowly escaping gas, according to the method of Thomson and Joule. If we introduce t (of § 160) instead of T into equation (86), which represents the theory of those experiments on the absolute temperature scale, we have

$$\Delta T = \frac{dT}{dt} \Delta t$$

$$\left(\frac{\partial v}{\partial T}\right)_p = \left(\frac{\partial v}{\partial t}\right)_p \cdot \frac{dt}{dT}$$

$$c_p = \left(\frac{q}{dT}\right)_p = \left(\frac{q}{dt}\right)_p \cdot \frac{dt}{dT} = c_p' \frac{dt}{dT},$$

where c_p' is the specific heat at constant pressure, determined by a t thermometer. Consequently, by (86),

$$\Delta t = \frac{T\left(\frac{\partial v}{\partial t}\right)_p \cdot \frac{dt}{dT} - v}{c_p'}\Delta p$$

and again, by integration,

$$\log \frac{T}{T_0} = \int_{t_0}^{t} \frac{\left(\frac{\delta v}{\delta t}\right)_p dt}{v + c_p'\frac{\Delta t}{\Delta p}} = J. \quad . \quad . \quad . \quad (90)$$

The expression to be integrated again contains quantities which may be measured directly with comparative ease.

§ 163. The stipulation of § 160, that, at the freezing-point of water, $T = T_0 = 273$, implies the knowledge of the coefficient of expansion, α, of perfect gases. Strictly speaking, however, all gases show at all temperatures deviations from the behaviour of perfect gases, and disagree with one another. To rid ourselves of any definite assumption about α, we return to our original definition of temperature, viz. that the difference between the absolute temperature of water boiling under atmospheric pressure (T_1), and that of water freezing under the same pressure (T_0), shall be

$$T_1 - T_0 = 100 \quad . \quad . \quad . \quad . \quad (91)$$

Now, if t_1 be the boiling-point of water, measured by means of a t thermometer, then, by (90),

$$\log \frac{T_1}{T_0} = \int_{t_0}^{t_1} \frac{\left(\frac{\partial v}{\partial t}\right)_p dt}{v + c_p'\frac{\Delta t}{\Delta p}} = J_1 \quad . \quad . \quad . \quad (92)$$

and, eliminating T_0 and T_1 from (90), (91), and (92), we find the absolute temperature :

$$T = \frac{100e^J}{e^{J_1} - 1}. \quad . \quad . \quad . \quad . \quad (93)$$

From this we obtain the coefficient of thermal expansion of a perfect gas, independently of any gas thermometer,

$$\alpha = \frac{1}{T_0} = \frac{e^{J_1} - 1}{100}. \quad . \quad . \quad . \quad (94)$$

Since, in both J and J_1, the expression to be integrated depends necessarily on t only, it is sufficient for the calculation of the value of the integral to experiment at different temperatures under some simplifying condition, as, for instance, always at the same pressure (atmospheric pressure).

§ **164.** The formula may be still further simplified by using as thermometric substance in the t thermometer the same gas as that used in Thomson and Joule's experiments. The coefficient of expansion, α', referred to temperature t, is then a constant, and if, as is usual, we put $t_0 = 0$, and $t_1 = 100$,

$$v = v_0(1 + \alpha't),$$

v_0 being the specific volume at the melting-point of ice under atmospheric pressure. Also

$$\left(\frac{\partial v}{\partial t}\right)_p = \alpha'v_0,$$

Hence, by (90),

$$J = \int_0^t \frac{\alpha'dt}{1 + \alpha't + \frac{c_p'}{v_0} \cdot \frac{\Delta t}{\Delta p}},$$

and, by (92),

$$J_1 = \int_0^{100} \frac{\alpha'dt}{1 + \alpha't + \frac{c_p'}{v_0} \cdot \frac{\Delta t}{\Delta p}}.$$

In the case of an almost perfect gas (*e.g.* air), Δt is small, and the term $\dfrac{c_p'}{v_0} \cdot \dfrac{\Delta t}{\Delta p}$ acts merely as a correction term, and, therefore, no great degree of accuracy is required in the determination of c_p' and v_0. For a perfect gas we should have $\Delta t = 0$, and, from the last two equations,

$$J = \log(1 + \alpha't), \quad J_1 = \log(1 + 100\alpha');$$

therefore, by (93),

$$T = t + \frac{1}{\alpha''},$$

and, by (94),

$$\alpha = \frac{1}{T_0} = \alpha',$$

as it should be.

As soon as accurate measurement of even a single substance has determined T as a function of t, the question regarding the value of the absolute temperature may be considered as solved for all cases.

The absolute temperature may be determined not only by experiments on homogeneous substances, but also by application of the second law to heterogeneous systems (cf. § 177).

§ **164**a. Although the second law of thermodynamics is so important in principle and so indispensable at extreme temperatures, for fixing the absolute scale of temperature, yet perhaps, up to now, the most accurate value of α, the coefficient of expansion of a perfect gas, has been deduced, not from the second law, but direct from the coefficient of expansion of real gases. This has been done by using D. Berthelot's (§ 25) modification of van der Waals' equation, which is the more accurate, the further the gas is removed from the point of condensation. In this way, the coefficient of expansion α of a perfect gas is found to be 0·0036618, and accordingly the absolute temperature of the freezing-point of water

$$T_0 = \frac{1}{\alpha} = 273 \cdot 09.$$

CHAPTER II.

SYSTEM IN DIFFERENT STATES OF AGGREGATION.

§ 165. WE shall discuss in this chapter the equilibrium of a system which may consist of solid, liquid, and gaseous portions. We assume that the state of each of these portions is fully determined by mass, temperature, and volume; or, in other words, that the system is formed of but one independent constituent (§ 198). For this it is not necessary that any portion of the system should be chemically homogeneous. Indeed, the question with regard to the chemical homogeneity cannot, in general, be completely answered (§ 92). It is still very uncertain whether the molecules of liquid water are the same as those of ice. In fact, the anomalous properties of water in the neighbourhood of its freezing-point make it probable that even in the liquid state its molecules are of different kinds. The decision of such questions has no bearing on the investigations of this chapter. The system may even consist of a mixture of substances in any proportion; that is, it may be a solution or an alloy. What we assume is only this : that the state of each of its homogeneous portions is quite definite when the temperature T and the specific volume v are definitely given, and that, if the system consists of different substances, their proportion is the same in all portions of the system. We may now enunciate our problem in the following manner :—

Let us imagine a substance of given total mass, M, enclosed in a receptacle of volume, V, and the energy, U, added to it by heat-conduction. If the system be now isolated and left to itself, M, V, and U will remain constant, while the entropy, Φ, will increase. We shall now investigate the state or states of equilibrium which the system may assume, finding at the

139

same time the conditions of its stability or instability. This investigation may be completely carried through by means of the proposition expressed in equation (77), that of all the states that may adiabatically arise from one another, the most stable state of equilibrium is characterized by an absolute maximum of the entropy. The entropy may in general, however, as we shall see, assume several relative maxima, under the given external conditions. Each maximum, which is not the absolute one, will correspond to a more or less unstable equilibrium. The system in a state of this kind (*e.g.* as supersaturated vapour) may occasionally, upon appropriate, very slight disturbances, undergo a finite change, and pass into another state of equilibrium, which necessarily corresponds to a greater value of the entropy.

§ **166.** We have now to find, first of all, the states in which the entropy Φ becomes a maximum. The most general assumption regarding the state of the system is that it consists of a solid, a liquid, and a gaseous portion. Denoting the masses of these portions by M_1, M_2, M_3, but leaving open, for the present, the question as to which particular portion each suffix refers, we have for the entire mass of the system

$$M_1 + M_2 + M_3 = M.$$

All the quantities are positive, but some may be zero. Further, since the state under discussion is to be one of equilibrium, each portion of the system, also when taken alone, must be in equilibrium, and therefore of uniform temperature and density. To each of them, therefore, we may apply the propositions which were deduced in the preceding chapter for homogeneous substances. If v_1, v_2, v_3, denote the specific volumes, the given volume of the system is

$$M_1v_1 + M_2v_2 + M_3v_3 = V.$$

Similarly, the given energy is

$$M_1u_1 + M_2u_2 + M_3u_3 = U,$$

where u_1, u_2, u_3, denote the specific energies of the portions.

These three equations represent the given *external* conditions.

§ 167. For the entropy of the system we have

$$\Phi = M_1\phi_1 + M_2\phi_2 + M_3\phi_3,$$

ϕ_1, ϕ_2, ϕ_3, being the specific entropies.

For an infinitesimal change of state this equation gives

$$\delta\Phi = \Sigma M_1\delta\phi_1 + \Sigma\phi_1\delta M_1.$$

Since, by (61), we have, in general,

$$\delta\phi = \frac{\delta u + p\delta v}{T},$$

we obtain

$$\delta\Phi = \Sigma\frac{M_1\delta u_1}{T_1} + \Sigma\frac{M_1 p_1\delta v_1}{T_1} + \Sigma\phi_1\delta M_1 \quad . \quad (95)$$

These variations are not all independent of one another. In fact, from the equations of the imposed (external) conditions, it follows that

$$\left. \begin{array}{c} \Sigma\delta M_1 = 0 \\ \Sigma M_1\delta v_1 + \Sigma v_1\delta M_1 = 0 \\ \Sigma M_1\delta u_1 + \Sigma u_1\delta M_1 = 0 \end{array} \right\} \quad . \quad . \quad . \quad (96)$$

With the help of these equations we must eliminate from (95) any three variations, in order that it may contain only independent variations. If we substitute in (95), for instance, the values for δM_2, δv_2, and δu_2 taken from (96), the equation for $\delta\Phi$ becomes

$$\left. \begin{array}{l} \delta\Phi = \left(\dfrac{1}{T_1} - \dfrac{1}{T_2}\right)M_1\delta u_1 - \left(\dfrac{1}{T_2} - \dfrac{1}{T_3}\right)M_3\delta u_3 \\[2mm] + \left(\dfrac{p_1}{T_1} - \dfrac{p_2}{T_2}\right)M_1\delta v_1 - \left(\dfrac{p_2}{T_2} - \dfrac{p_3}{T_3}\right)M_3\delta v_3 \\[2mm] + \left(\phi_1 - \phi_2 - \dfrac{u_1 - u_2}{T_2} - \dfrac{p_2(v_1 - v_2)}{T_2}\right)\delta M_1 \\[2mm] - \left(\phi_2 - \phi_3 - \dfrac{u_2 - u_3}{T_2} - \dfrac{p_2(v_2 - v_3)}{T_2}\right)\delta M_3 \end{array} \right\} \quad (97)$$

Since the six variations occurring in this expression are now independent of one another, it is necessary that each of their six coefficients should vanish, in order that $\delta\Phi$ may be zero for all changes of state. Therefore

$$\left.\begin{array}{l} T_1 = T_2 = T_3 \,(= T) \\[4pt] p_1 = p_2 = p_3 \\[4pt] \phi_1 - \phi_2 = \dfrac{(u_1 - u_2) + p_1(v_1 - v_2)}{T} \\[10pt] \phi_2 - \phi_3 = \dfrac{u_2 - u_3 + p_2(v_2 - v_3)}{T} \end{array}\right\} \quad . \quad . \quad (98)$$

These six equations represent necessary properties of any state, which corresponds to a maximum value of the entropy, *i.e.* of any state of equilibrium. As the first four refer to equality of temperature and pressure, the main interest centres in the last two, which contain the thermodynamical theory of fusion, evaporation, and sublimation.

§ 168. These two equations may be considerably simplified by substituting the value of the specific entropy ϕ, which, as well as u and p, is here considered as a function of T and v. For, since (61) gives, in general,

$$d\phi = \frac{du + p\,dv}{T},$$

we get, by integration,

$$\phi_1 - \phi_2 = \int_2^1 \frac{du + p\,dv}{T},$$

where the upper limit of the integral is characterized by the values T_1, v_1, the lower by T_2, v_2. The path of integration is arbitrary, and does not influence the value of $\phi_1 - \phi_2$. Since, now, $T_1 = T_2 = T$ (by 98), we may select an isothermal path of integration (T const.). This gives

$$\phi_1 - \phi_2 = \frac{u_1 - u_2}{T} + \frac{1}{T}\int_{v_2}^{v_1} p\,dv.$$

The integration is to be taken along an isotherm, since p is a known function of T and v determined by the characteristic equation of the substance. Substituting the value of $\phi_1 - \phi_2$ in the equations (98), we have the relations :

$$\left.\begin{array}{c} \int_{v_2}^{v_1} p\, dv = p_1(v_1 - v_2) \\[2mm] \int_{v_3}^{v_2} p\, dv = p_2(v_2 - v_3) \\[2mm] p_1 = p_2 = p_3 \end{array}\right\} \quad \cdot \quad \cdot \quad \cdot \quad (99)$$

to which we add

With the four unknowns T, v_1, v_2, v_3, we have four equations which the state of equilibrium must satisfy. The constants which occur in these equations depend obviously only on the chemical nature of the substance, and in no way on the given values of the mass, M, the volume, V, and the energy, U, of the system. The equations (99) might therefore be called the system's *internal* or *intrinsic* conditions of equilibrium, while those of § 166 represent the *external* conditions imposed on the system.

§ **169.** Before discussing the values which the equations (99) give to the unknowns, we shall investigate generally whether, and under what condition, they lead to a maximum value of the entropy and not to a minimum value. It is necessary, for this purpose, to find the value of $\delta^2\Phi$. If this be negative for all virtual changes, then the state considered is certainly one of maximum entropy.

From the expression for $\delta\Phi$ (97) we obtain $\delta^2\Phi$, which may be greatly simplified with the help of the equations (98). The equations of the imposed external conditions, and the equations (96) further simplify the result, and we obtain, finally,

$$\delta^2\Phi = - \sum \frac{M_1 \delta\phi_1 \delta T_1}{T_1} + \sum \frac{M_1 \delta p_1 \delta v_1}{T_1}.$$

This may be written

$$T\delta^2\Phi = - \Sigma M_1(\delta\phi_1 \delta T_1 - \delta p_1 \delta v_1).$$

To reduce all variations to those of the independent variables, T and v, we may write, according to (81),

$$\delta\phi = \left(\frac{\partial\phi}{\partial T}\right)_v \delta T + \left(\frac{\partial\phi}{\partial v}\right)_T \delta v = \frac{c_v}{T}\delta T + \left(\frac{\partial p}{\partial T}\right)_v \delta v$$

and $\quad \delta p = \left(\frac{\partial p}{\partial T}\right)_v \delta T + \left(\frac{\partial p}{\partial v}\right)_T \delta v,$

$$\therefore \quad T\delta^2\Phi = -\sum M_1\left(\frac{(c_v)_1}{T}\delta T_1{}^2 - \left(\frac{\partial p_1}{\partial v_1}\right)_T \delta v_1{}^2\right). \quad . \quad . \quad (100)$$

Obviously, if the quantities $(c_v)_1$, $(c_v)_2$, $(c_v)_3$ be all positive, and the quantities $\left(\frac{\partial p_1}{\partial v}\right)_T \ldots$ all negative, $\delta^2\Phi$ is negative in all cases, and Φ is really a maximum, and the corresponding state is a state of equilibrium. Since c_v is the specific heat at constant volume, and therefore always positive, the condition of equilibrium depends on whether $\left(\frac{\partial p}{\partial v}\right)_T$ is negative for all three portions of the system or not. In the latter case there is no equilibrium. Experience immediately shows, however, that in any state of equilibrium $\frac{\partial p}{\partial v}$ is negative, since the pressure, whether positive or negative, and the volume always change in opposite directions. A glance at the graphical representation of p, as an isothermal function of v (Fig. 1, § 26), shows that there are certain states of the system in which $\frac{\partial p}{\partial v}$ is positive. These, however, can never be states of equilibrium, and are, therefore, not accessible to direct observation. If, on the other hand, $\frac{\partial p}{\partial v}$ be negative, it is a state of equilibrium, yet it need not be stable; for another state of equilibrium may be found to exist which corresponds to a greater value of the entropy.

We shall now discuss the values of the unknowns, T, v_1, v_2, v_3, which represent solutions of the conditions of equilibrium (98) or (99). Several such systems may be found. Thereafter, we shall deal (beginning at § 189) with the further question as to which of the different solutions in each case represents the most stable equilibrium under the given external conditions; *i.e.* which one leads to the largest value of the entropy of the system.

§ 170. First Solution.—If we put, in the first place,

$$v_1 = v_2 = v_3(= v)$$

all the equations (98) are satisfied, for, since the temperature is common to all three portions of the system, their states become absolutely identical. The entire system is, therefore, homogeneous. The state of the system is determined by the equations of § 166, which give the imposed conditions. In this case they are

$$M_1 + M_2 + M_3 = M$$
$$v(M_1 + M_2 + M_3) = V$$
$$u(M_1 + M_2 + M_3) = U$$
$$\therefore \quad v = \frac{V}{M} \text{ and } u = \frac{U}{M}.$$

From v and u, T may be found, since u was assumed to be a known function of T and v.

This solution has always a definite meaning; but, as we saw in equation (100), it represents a state of equilibrium only when $\dfrac{\partial p}{\partial v}$ is negative. If this be the case, then, the equilibrium is stable or unstable, according as under the external conditions there exists a state of greater entropy or not. This will be discussed later (§ 189).

§ 171. Second Solution.—If, in the second case, we put

$$v_1 \neq v_2, \ v_2 = v_3,$$

the states 2 and 3 coincide, and the equations (98) reduce to

$$\left. \begin{aligned} p_1 &= p_2 \\ \phi_1 - \phi_2 &= \frac{u_1 - u_2 + p_1(v_1 - v_2)}{T} \end{aligned} \right\} \quad . \quad (101)$$

or, instead of the second of these equations,

$$\int_{v_2}^{v_1} p \, dv = p_1(v_1 - v_2) \quad . \quad . \quad . \quad . \quad (102)$$

In this case two states of the system coexist; for instance, the gaseous and the liquid. The equations (101) contain three unknowns, T, v_1, v_2; and hence may serve to express v_1 and v_2, consequently also the pressure $p_1 = p_2$, and the specific energies u_1 and u_2, as definite functions of the temperature T. The internal state of two heterogeneous portions of the same substance in contact with one another is, therefore, completely determined by the temperature. The temperature, as well as the masses of the two portions, may be found from the imposed conditions (§ 166), which are, in this case,

$$\left. \begin{array}{l} M_1 + (M_2 + M_3) = M \\ M_1 v_1 + (M_2 + M_3)v_2 = V \\ M_1 u_1 + (M_2 + M_3)u_2 = U \end{array} \right\} \quad . \quad . \quad (103)$$

These three equations serve for the determination of the three last unknowns, T, M_1, and $M_2 + M_3$. This completely determines the physical state, for, in the case of the masses M_2 and M_3, it is obviously sufficient to know their sum. Of course, the result can only bear a physical interpretation if both M_1 and $M_2 + M_3$ have positive values.

§ 172. An examination of equation (102) shows that it can be satisfied only if the pressure, p, which is known to have the same value ($p_1 = p_2$) for both limits of the integral, assume between the limits values which are partly larger and partly smaller than p_1. Some of these, then, must correspond to unstable states (§ 169), since in certain places p and v increase simultaneously $\left(\dfrac{\partial p}{\partial v} > 0\right)$. The equation admits of a simple geometrical interpretation with the help of the above-mentioned graphical representation of the characteristic equation by isotherms (Fig. 1, § 26). For the integral $\int_2^1 p \, dv$ is represented by the area bounded by the isotherm, the axis of abscissæ, and the ordinates at v_1 and v_2, while the product $p_1(v_1 - v_2)$ is the rectangle formed by the same ordinates ($p_1 = p_2$), and the length ($v_1 - v_2$).

We learn, therefore, from equation (102) that in every isotherm the pressure, under which two states of aggregation of the substance may be kept in lasting contact, is represented by the ordinate of the straight line parallel to the axis of abscissæ, which intercepts equal areas on both sides of the isotherm. Such a line is represented by ABC in Fig. 1. We are thus enabled to deduce directly from the characteristic equation for homogeneous, stable and unstable, states the functional relation between the pressure, the density of the saturated vapour and of the liquid in contact with it, and the temperature.

Taking Clausius' equation (12a) as an empirical expression of the facts, we have, for the specific volume v_1 of the saturated vapour, and v_2 of the liquid in contact with it, the two conditions

$$\frac{RT}{v_1 - a} - \frac{c}{T(v_1 + b)^2} = \frac{RT}{v_2 - a} - \frac{c}{T(v_2 + b)^2},$$

and, from (102),

$$RT \log \frac{v_1 - a}{v_2 - a} - \frac{c}{T}\left(\frac{1}{v_2 + b} - \frac{1}{v_1 + b}\right) = (v_1 - v_2)\left(\frac{RT}{v_1 - a} - \frac{c}{T(v_1 + b)^2}\right).$$

By means of these v_1, v_2, and $p_1 = p_2$ may be expressed as functions of T, or, still more conveniently, v_1, v_2, p_1, and T as functions of some appropriately selected independent variable. At the critical temperature it gives $v_1 = v_2$ (§ 30a). At higher temperatures v_1 and v_2 are imaginary. At infinitely low temperatures $v_1 = \infty$, $v_2 = a$.[*]

With Clausius' values of the constants for carbon dioxide (§ 25), this calculation furnishes results which show a satisfactory agreement with Andrews' observations. According to Thiesen, however, Clausius' equation possesses no general significance.

§ 173. The import of equations (101) can be more simply expressed if, instead of the entropy, we introduce the free

[*] *Wied. Ann.*, **13**, p. 535, 1881.

energy F, and still simpler if we use the function ψ. We shall here keep to the free energy. This is by (71)

$$f = u - T\phi \quad . \quad . \quad . \quad . \quad (104)$$

per unit mass.

Then the equations (101) become, simply,

$$p_1 = p_2 \quad . \quad . \quad . \quad . \quad . \quad (105)$$

$$f_2 - f_1 = p_1(v_1 - v_2) \quad . \quad . \quad . \quad (106)$$

The function f satisfies the following simple conditions.

By (79a),
$$\left(\frac{\partial f}{\partial T}\right)_v = -\phi. \quad . \quad . \quad . \quad . \quad (107)$$

and
$$\left(\frac{\partial f}{\partial v}\right)_T = -p. \quad . \quad . \quad . \quad . \quad (108)$$

The conditions of equilibrium for two states of aggregation in mutual contact hold for the three possible combinations of the solid and liquid, liquid and gaseous, gaseous and solid states. In order to fix our ideas, however, we shall discuss that solution of those equations which corresponds to the contact of a liquid with its vapour. Denoting the vapour by the subscript 1, the liquid by 2, v_1 is then the specific volume of the saturated vapour at the temperature T; $p_1 = p_2$, its pressure; v_2 the specific volume of the liquid with which it is in contact. All these quantities, then, are functions of the temperature only, which agrees with experience.

§ **174.** Further theorems may be arrived at by the differentiation of the conditions of equilibrium with respect to T. Since all variables now depend only on T, we shall use $\frac{d}{dT}$ to indicate this total differentiation, while partial differentiation with respect to T will be expressed, as hitherto, by $\frac{\partial}{\partial T}$.

Equations (105) and (106), thus differentiated, give

$$\frac{dp_1}{dT} = \frac{dp_2}{dT}$$

and $$\frac{df_2}{dT} - \frac{df_1}{dT} = (v_1 - v_2)\frac{dp_1}{dT} + p_1\left(\frac{dv_1}{dT} - \frac{dv_2}{dT}\right).$$

But, by (107) and (108), we have

$$\frac{df_2}{dT} - \frac{df_1}{dT} = \left(\frac{\partial f_2}{\partial T}\right)_v + \left(\frac{\partial f_2}{\partial v}\right)_T \cdot \frac{dv_2}{dT} - \left(\frac{\partial f_1}{\partial T}\right)_v - \left(\frac{\partial f_1}{\partial v}\right)_T \cdot \frac{dv_1}{dT}$$

$$= -\phi_2 - p_2\frac{dv_2}{dT} + \phi_1 + p_1\frac{dv_1}{dT},$$

whence, by substitution,

$$\phi_1 - \phi_2 = (v_1 - v_2)\frac{dp_1}{dT},$$

or, finally, by (101),

$$(u_1 - u_2) + p_1(v_1 - v_2) = T(v_1 - v_2)\frac{dp_1}{dT}. \quad (109)$$

Here the left-hand side of the equation, according to equation (17) of the first law, represents the heat of vaporization, L, of the liquid. It is the heat which must be added to unit mass of the liquid, in order to completely change it to vapour under the constant pressure of its saturated vapour. For the corresponding change of energy is $u_1 - u_2$, and the external work performed, here negative, amounts to

$$W = -p_1(v_1 - v_2)$$
$$\therefore L = u_1 - u_2 + p_1(v_1 - v_2), \quad . \quad . \quad (110)$$

whence $$L = T(v_1 - v_2)\frac{dp_1}{dT} \quad . \quad . \quad . \quad . \quad (111)$$

This equation, deduced by Clapeyron from Carnot's theory, but first rigorously proved by Clausius, may be used for the determination of the heat of vaporization at any temperature, if we know the specific volumes of the saturated vapour and the liquid, as well as the relation between the pressure of the saturated vapour and the temperature. This formula has been verified by experiment in a large number of cases.

§ **175.** As an example, we shall calculate the heat of vaporization of water at $100°$ C., *i.e.* under atmospheric pressure, from the following data :—

T $= 273 + 100 = 373$.

$v_1 = 1674$ (volume of 1 gr. of saturated water vapour at 100° in c.c., according to Knoblauch, Linde and Klebe).

$v_2 = 1$ (volume of 1 gr. of water at 100° in c.c.).

$\frac{dp_1}{dT}$ is found from the experiments of Holborn and Henning.

Saturated water vapour at 100° C. gave an increase of pressure of 27·12 mm. of mercury for a rise of 1° C. In absolute units, by § 7,

$$\frac{dp_1}{dT} = \frac{27·12}{760} \times 1013250.$$

Thus, the required latent heat of vaporization is

$$L = \frac{373 \times 1673 \times 27·12 \times 1013250}{760 \times 4·19 \times 10^7} = 539 \; \frac{cal.}{grm.}$$

Henning found the latent heat of vaporisation of water at 100° C. by direct measurement to be 538·7 $\frac{cal.}{grm.}$, in excellent agreement with the calculated value.

§ 176. As equation (110) shows, part of the heat of vaporization, L, corresponds to an increase of energy, and part to external work. To find the relation between these two it is most convenient to find the ratio of the external work to the latent heat of vaporization, viz.

$$\frac{p_1(v_1 - v_2)}{L} = \frac{p_1}{T\frac{dp_1}{dT}}.$$

In the above case $p_1 = 760$ mm., T = 373, $\frac{dp_1}{dT} = 27·12$ mm., and therefore,

$$\frac{p_1(v_1 - v_2)}{L} = \frac{760}{373 \times 27·12} = 0·075.$$

This shows that the external work forms only a small part of the value of the latent heat of vaporization.

§ 177. Equation (111) also leads to a method of calculating the absolute temperature T, when the latent heat of vaporization, L, as well as the pressure and the density of the saturated vapour and the liquid, have been determined by experiment in terms of any scale of temperature t (§ 160). We have

$$L = T(v_1 - v_2)\frac{dp_1}{dt} \cdot \frac{dt}{dT},$$

$$\therefore \log T = \int \frac{v_1 - v_2}{L} \cdot \frac{dp_1}{dt} dt,$$

and therefore T may be determined as a function of t. It is obvious that any equation between measurable quantities, deduced from the second law, may be utilized for a determination of the absolute temperature. The question as to which of those methods deserves preference is to be decided by the degree of accuracy to be obtained in the actual measurements.

§ 178. A simple approximation formula, which in many cases gives good, though in some, only fair results, may be obtained by neglecting in the equation (111) the specific volume of the liquid, v_2, in comparison with that of the vapour, v_1, and assuming for the vapour the characteristic equation of a perfect gas. Then, by (14),

$$v_1 = \frac{R}{m} \cdot \frac{T}{p_1},$$

where R is the absolute gas constant, and m the molecular weight of the vapour. Equation (111) then becomes

$$L = \frac{R}{m} \cdot \frac{T^2}{p_1} \cdot \frac{dp_1}{dT} \quad . \quad . \quad . \quad . \quad (112)$$

For water at 100° C. we have R = 1·985; $m = H_2O = 18$; T = 373; $p_1 = 760$ mm.; $\frac{dp_1}{dT} = 27\cdot12$ mm. Hence the latent heat of vaporization is

$$L = \frac{1\cdot985 \times 373^2 \times 27\cdot12}{18 \times 760} = 547\cdot5 \frac{cal.}{grm.}$$

This value is somewhat large (§ 175). The cause of this lies in the fact that the volume of saturated water vapour at 100° is in reality smaller than that calculated from the characteristic equation of a perfect gas of molecular weight 18. But, for this very reason, accurate measurement of the heat of vaporization may serve as a means of estimating from the second law the amount by which the density of a vapour deviates from the ideal value.

Another kind of approximation formula, valid within the same limits, is found by substituting in (109) the value of the specific energy $u_1 = c_v T + $ const., which, by (39), holds for perfect gases. We may put the specific energy of the liquid $u_2 = c_2 T + $ const., if we assume its specific heat, c_2, to be constant, and neglect the external work. It then follows from (109) that

$$(c_v - c_2)T + \text{const.} + \frac{RT}{m} = \frac{R}{m} \cdot \frac{T^2}{p_1} \cdot \frac{dp_1}{dT}.$$

If we multiply both sides by $\frac{dT}{T^2}$, this equation may be integrated, term by term, and we finally obtain with the help of (33)

$$p_1 = ae^{-\frac{b}{T}} \cdot T^{\frac{m}{R}(c_p - c_2)}$$

where a and b are positive constants; c_p and c_2 the specific heats of the vapour and the liquid, at constant pressure. This equation gives a relation between the pressure of a saturated vapour and its temperature.

H. Hertz has calculated the constants for mercury.* It must, however, be noted that at low temperatures the assumption that c_2 is constant is no longer correct (see §§ 284 and 288).

§ **179.** Equation (111) is applicable to the processes of fusion and sublimation in the same manner as to that of evaporation. In the first case L denotes the latent heat of

---

* *Wied. Ann. d. Phys.,* **17**, p. 193, 1882.

fusion of the substance, if the subscript 1 correspond to the liquid state and 2 to the solid state, and p_1 the melting pressure, *i.e.* the pressure under which the solid and the liquid substance may be in contact and in equilibrium. The melting pressure, therefore, just as the pressure of evaporation, depends on the temperature only. Conversely, a change of pressure produces a change in the melting point:

$$\frac{d\mathrm{T}}{dp_1} = \frac{\mathrm{T}(v_1 - v_2)}{\mathrm{L}} \quad . \quad . \quad . \quad (113)$$

For ice at $0°$ C. and under atmospheric pressure, we have

$\mathrm{L} = 80 \times 4{\cdot}19 \times 10^7$ (heat of fusion of 1 gr. of ice in c.g.s. units);

$\mathrm{T} = 273;$

$v_1 = 1{\cdot}000$ (vol. of 1 gr. of water at $0°$ C. in c.c.);

$v_2 = 1{\cdot}091$ (vol. of 1 gr. of ice at $0°$ C. in c.c.).

To obtain $\dfrac{d\mathrm{T}}{dp_1}$ in atmospheres we must multiply by 1,013,250:

$$\frac{d\mathrm{T}}{dp_1} = -\frac{273 \times 0{\cdot}091 \times 1013250}{80 \times 4{\cdot}19 \times 10^7} = -0{\cdot}0075. \quad (114)$$

On increasing the external pressure by 1 atmosphere, the melting point of ice will, therefore, be lowered by $0{\cdot}0075°$ C.; or, to lower the melting point of ice by $1°$ C., the pressure must be increased by about 130 atmospheres. This was first verified by the measurements of W. Thomson (Lord Kelvin). Equation (113) shows that, conversely, the melting point of substances, which expand on melting, is raised by an increase of pressure. This has been qualitatively and quantitatively verified by experiment.

§ **180.** By means of the equations (101) still further important properties of substances in different states may be shown to depend on one another. From these, along with (110), we obtain

$$\frac{\mathrm{L}}{\mathrm{T}} = \phi_1 - \phi_2.$$

Differentiating this with respect to T, we have

$$\frac{1}{T} \cdot \frac{dL}{dT} - \frac{L}{T^2} = \left(\frac{\partial \phi_1}{\partial T}\right)_p + \left(\frac{\partial \phi_1}{\partial p}\right)_T \frac{dp_1}{dT} - \left(\frac{\partial \phi_2}{\partial T}\right)_p - \left(\frac{\partial \phi_2}{\partial p}\right)_T \frac{dp_2}{dT}$$

or by (84a) and (84b)

$$\frac{(c_p)_1}{T} - \left(\frac{\partial v_1}{\partial T}\right)_p \frac{dp_1}{dT} - \frac{(c_p)_2}{T} + \left(\frac{\partial v_2}{\partial T}\right)_p \frac{dp_2}{dT}.$$

Since, from (111),

$$\frac{dp_1}{dT} = \frac{dp_2}{dT} = \frac{L}{T(v_1 - v_2)},$$

we obtain, finally,

$$(c_p)_1 - (c_p)_2 = \frac{dL}{dT} - \frac{L}{T} + \frac{L}{v_1 - v_2} \cdot \left[\left(\frac{\partial v_1}{\partial T}\right)_p - \left(\frac{\partial v_2}{\partial T}\right)_p\right]. \quad (115)$$

This equation again leads to a test of the second law, since all the quantities in it may be measured independently of one another.

§ **181.** We shall again take as example saturated water vapour at 100° C. under atmospheric pressure, and calculate its specific heat at constant pressure. We have the following data:

$(c_p)_2 = 1\cdot01$ (spec. heat of liquid water at 100° C.);

$\quad$ L $= 539$ cal./grm. at 100° C.

$\quad$ T $= 373$;

$\quad$ Further *

$$\frac{dL}{dT} = -0\cdot64 \text{ (the decrease of the heat of vapor-}$$
$$\text{ization with rise of temperature)}$$

$$v_1 = 1674 \text{ (the volume of 1 gr. of saturated water}$$
$$\text{vapour at 100° C.)}$$

$$\left(\frac{\partial v_1}{\partial T}\right)_p = 4\cdot813 \text{ (the isobaric coefficient of expansion of}$$
$$\text{water vapour).}$$

* O. Knoblauch, R. Linde and H. Klebe, "Mitteilungen über Forschungs-arbeiten herausgegeben vom Verein Deutscher Ingenieure," Heft 21, Berlin, 1905. Harvey N. Davis, *Proc. of the American Academy of Arts and Science*, Vol. 45, p. 265, 1910. F. Henning, *Zeitschr. f. Physik.*, **2**, p. 197, 1920.

The corresponding values for liquid water are :

$$v_2 = 1 \cdot 0 \text{ c.c. per gr.}$$

$$\left(\frac{\partial v_2}{\partial \mathrm{T}}\right)_p = 0 \cdot 001.$$

These values substituted in (115) give

$$(c_p)_1 - (c_p)_2 = -0 \cdot 54,$$
$$(c_p)_1 = (c_p)_2 - 0 \cdot 54 = 1 \cdot 01 - 0 \cdot 54 = 0 \cdot 47 \text{ cal./grm. per } 1^\circ \text{ C.}$$

in agreement with direct measurement.

§ **182.** The relation (115) may be simplified, but is inaccurate, if we neglect the volume v_2 of the liquid in comparison with v_1, that of the vapour, and apply to the vapour the characteristic equation of a perfect gas.

Then
$$v_1 = \frac{\mathrm{R}\mathrm{T}}{mp_1},$$

$$\left(\frac{\partial v_1}{\partial \mathrm{T}}\right)_p = \frac{\mathrm{R}}{mp_1},$$

and equation (115) becomes, simply,

$$(c_p)_1 - (c_p)_2 = \frac{d\mathrm{L}}{d\mathrm{T}}.$$

In our example,
$$(c_p)_1 - (c_p)_2 = -0 \cdot 64$$
$$(c_p)_1 = 1 \cdot 01 - 0 \cdot 64 = 0 \cdot 37 \text{ cal./grm. per } 1^\circ \text{ C.}$$

a value considerably too small.

§ **183.** We shall now apply the relation (115) to the melting of ice at 0° C. and under atmospheric pressure. The subscript 1 now refers to the liquid state, and 2 to the solid state. The relation between the latent heat of fusion of ice and the temperature has probably never been measured. It may, however, be calculated from (115), which gives

$$\frac{d\mathrm{L}}{d\mathrm{T}} = (c_p)_1 - (c_p)_2 + \frac{\mathrm{L}}{\mathrm{T}} - \frac{\mathrm{L}}{v_1 - v_2}\left[\left(\frac{\partial v_1}{\partial \mathrm{T}}\right)_p - \left(\frac{\partial v_2}{\partial \mathrm{T}}\right)_p\right]$$

in which

$(c_p)_1 = 1$ (spec. heat of water at $0°$ C.);

$(c_p)_2 = 0.50$ (spec. heat of ice at $0°$ C.);

$\qquad$ L $= 80$; T $= 273$; $v_1 = 1$; $v_2 = 1.09$;

$\left(\dfrac{\partial v_1}{\partial T}\right)_p = -0.00006$ (thermal coeff. of expansion of water at

$\qquad 0°$ C.);

$\left(\dfrac{\partial v_2}{\partial T}\right) = 0.00011$ (thermal coeff. of expansion of ice at $0°$ C.).

Hence, by the above equation,

$$\frac{d\text{L}}{d\text{T}} = 0.66,$$

i.e. if the melting point of ice be lowered $1°$ C. by an appropriate increase of the external pressure, its heat of fusion decreases by 0.66 cal./grm.

§ **184.** It has been repeatedly mentioned in the early chapters, that, besides the specific heat at constant pressure, or constant volume, any number of specific heats may be defined according to the conditions under which the heating takes place. Equation (23) of the first law holds in each case :

$$c = \frac{du}{d\text{T}} + p\frac{dv}{d\text{T}}.$$

In the case of saturated vapours special interest attaches to the process of heating, which keeps them permanently in a state of saturation. Denoting by s_1 the specific heat of the vapour corresponding to this process (Clausius called it the specific heat of " the saturated vapour "), we have

$$s_1 = \frac{du_1}{d\text{T}} + p_1\frac{dv_1}{d\text{T}} \quad . \quad . \quad . \quad . \quad (116)$$

No off-hand statement can be made with regard to the value of s_1; even its sign must in the meantime remain uncertain. For, if during a rise of temperature of $1°$ the vapour is to remain just saturated, it must evidently be compressed while being heated, since the specific volume of the saturated vapour decreases as the temperature rises. This compression, how-ever, generates heat, and the question is, whether the latter is so considerable that it must be in part withdrawn by con-duction, so as not to superheat the vapour. Two cases may, therefore, arise : (1) The heat of compression may be con-siderable, and the withdrawal of heat is necessary to maintain saturation at the higher temperature, *i.e.* s_1 is negative. (2) The heat of compression may be too slight to prevent the compressed vapour, without the addition of heat, from becom-ing supersaturated. Then, s_1 has a positive value. Between the two there is a limiting case ($s_1 = 0$), where the heat of compression is exactly sufficient to maintain saturation. In this case the curve of the saturated vapour coincides with that of adiabatic compression. Watt assumed this to be the case for steam.

It is now easy to calculate s_1 from the above formulæ. Calling s_2 the corresponding specific heat of the liquid, we have

$$s_2 = \frac{du_2}{dT} + p_2\frac{dv_2}{dT} \quad . \quad . \quad . \quad (117)$$

During heating, the liquid is kept constantly under the pressure of its saturated vapour. Since the external pressure, unless it amounts to many atmospheres, has no appreciable influence on the state of a liquid, the value of s_2 practically coincides with that of the specific heat at constant pressure,

$$s_2 = (c_p)_2 \quad . \quad . \quad . \quad . \quad (118)$$

Subtracting (117) from (116), we get

$$s_1 - s_2 = \frac{d(u_1 - u_2)}{dT} + p_1\frac{d(v_1 - v_2)}{dT}.$$

But (110), differentiated with respect to T, gives

$$\frac{dL}{dT} = \frac{d(u_1 - u_2)}{dT} + p_1\frac{d(v_1 - v_2)}{dT} + (v_1 - v_2)\frac{dp_1}{dT},$$

$$\therefore \ s_1 - s_2 = \frac{dL}{dT} - (v_1 - v_2)\frac{dp_1}{dT},$$

or, by (118) and (111),

$$s_1 = (c_p)_2 + \frac{d\mathrm{L}}{d\mathrm{T}} - \frac{\mathrm{L}}{\mathrm{T}}.$$

For saturated water vapour at 100°, we have, as above,

$$(c_p)_2 = 1\cdot01;\ \frac{d\mathrm{L}}{d\mathrm{T}} = -\ 0\cdot64;\ \mathrm{L} = 539;\ \mathrm{T} = 373;$$

whence

$$s_1 = 1\cdot01 - 0\cdot64 - 1\cdot44 = -\ 1\cdot07.$$

Water vapour at 100° C. represents the first of the cases described above, *i.e.* saturated water vapour at 100° is super-heated by adiabatic compression. Conversely, saturated water vapour at 100° becomes supersaturated by adiabatic expansion. The influence of the heat of compression (or expansion) is greater than the influence of the increase (or decrease) of the density. Some other vapours behave in the opposite way.

§ **185.** It may happen that, for a given value of T, the values of v_1 and v_2, which are fully determined by the equation (101), become equal. Then the two states which are in contact with one another are identical. Such a value of T is called a *critical temperature* of the substance. From a purely mathematical point of view, every substance must be supposed to have a critical temperature for each of the three combinations, solid-liquid, liquid-gas, gas-solid. This critical temperature, however, will not always be real. The critical temperature T and the critical volume $v_1 = v_2$, fully determine the critical state. We may calculate it from the equations (101) by finding the condition that $v_1 - v_2$ should vanish. If we first assume $v_1 - v_2$ to be very small, Taylor's theorem then gives for any volume v, lying between v_1 and v_2,

$$p = p_2 + \left(\frac{\partial p}{\partial v}\right)_2 (v - v_2) + \tfrac{1}{2}\left(\frac{\partial^2 p}{\partial v^2}\right)_2 (v - v_2)^2 . \quad . \quad (119)$$

and therefore the first equation (101) becomes

$$p_2 + \left(\frac{\partial p}{\partial v}\right)_2 (v_1 - v_2) + \tfrac{1}{2}\left(\frac{\partial^2 p}{\partial v^2}\right)_2 (v_1 - v_2)^2 = p_2,$$

and equation (102), by the integration of (119) with respect to v, gives

$$p_2(v_1 - v_2) + \tfrac{1}{2}\Big(\frac{\partial p}{\partial v}\Big)_2 (v_1 - v_2)^2 + \tfrac{1}{2 \cdot 3}\Big(\frac{\partial^2 p}{\partial v^2}\Big)_2 (v_1 - v_2)^3 = p_2(v_1 - v_2).$$

The last two equations give, as the conditions of the critical state,

$$\Big(\frac{\partial p}{\partial v}\Big)_2 = 0, \text{ and } \Big(\frac{\partial^2 p}{\partial v^2}\Big)_2 = 0.$$

These conditions agree with those found in § 30. They are there geometrically illustrated by the curve of the critical isotherm. In the critical state the compressibility is infinite; so are also the thermal coefficient of expansion and the specific heat at constant pressure; the heat of vaporization is zero.

At all temperatures other than the critical one, the values of v_1 and v_2 are different. On one side of the critical isotherm they have real, on the other imaginary values. In this latter case our solution of the problem of equilibrium no longer admits of a physical interpretation.

§ 186. Third Solution.—In the third place, we shall assume that in the conditions of equilibrium (98)

$$v_1 \neq v_2 \neq v_3.$$

We have then, without further simplification,

$$\left.\begin{aligned}
p_1 &= p_2 = p_3 \\
\phi_1 - \phi_2 &= \frac{u_1 - u_2 + p_1(v_1 - v_2)}{T} \\
\phi_2 - \phi_3 &= \frac{u_2 - u_3 + p_1(v_2 - v_3)}{T}
\end{aligned}\right\} \quad . \quad . \quad (120)$$

These refer to a state in which the three states of aggregation are simultaneously present. There are four equations, and these assign definite values to the four unknowns T, v_1, v_2, v_3. The coexistence of the three states of aggregation in equilibrium is, therefore, possible only at a definite temperature, and with definite densities; therefore, also, at a definite pressure. We shall call this temperature the *funda-mental temperature*, and the corresponding pressure the *fundamental pressure* of the substance. According to equations (120), the fundamental temperature is characterized by the condition that at it the pressure of the saturated vapour is equal to the pressure of fusion. It necessarily follows, by addition of the last two equations, that this pressure is also equal to the pressure of sublimation.

After the fundamental temperature and pressure have been found, the external conditions of § 166—

$$\left.\begin{array}{l} M_1 + M_2 + M_3 = M \\ M_1v_1 + M_2v_2 + M_3v_3 = V \\ M_1u_1 + M_2u_2 + M_3u_3 = U \end{array}\right\} \quad . \quad . \quad (121)$$

uniquely determine the masses of the three portions of the substance. The solution, however, can be interpreted physically only if M_1, M_2, and M_3 are positive.

§ 187. Let us determine, *e.g.*, the fundamental state of water. 0° C. is not its fundamental temperature, for at 0° C. the maximum vapour pressure of water is 4·58 mm., but the melting pressure of ice is 760 mm. Now, the latter decreases with rise of temperature, while the maximum vapour pressure increases. A coincidence of the two is, therefore, to be expected at a temperature somewhat higher than 0° C. According to equation (114), the melting point of ice rises by 0·0075° C. approximately, when the pressure is lowered from 760 mm. to 4·58 mm. The fundamental temperature of water is, then, approximately, 0·0075° C. At this temperature the maximum vapour pressure of water nearly coincides with the melting pressure of ice, and, therefore, also with the maximum vapour pressure of ice. The specific volumes of water in the three states are, therefore,

$$v_1 = 206{,}000; \quad v_2 = 1; \quad v_3 = 1·09 \text{ c.c.}$$

At all temperatures other than the fundamental temperature, the pressures of vaporization, of fusion, and of sublimation differ from one another.

§ **188.** We return once more to the intrinsic conditions of equilibrium (101) which hold for each of the three combinations of two states of aggregation. The pressure p_1, and the specific volumes of the two portions of the substance, in each case depend only on the temperature, and are determined by (101). It is necessary, however, to distinguish whether the saturated vapour is in contact with the liquid or the solid, since in these two cases the functions which express the pressure and specific volume in terms of the temperature are quite different. The state of the saturated vapour is determined only when there is given, besides the temperature, the state of aggregation with which it is in contact, whether it is in contact with the liquid or solid. The same applies to the other two states of aggregation. If we henceforth use the suffixes 1, 2, 3, in this order, to refer to the gaseous, liquid, and solid states, we shall be obliged to use two of them when we refer to a portion of the substance in a state of saturation. The first of these will refer to the state of the portion considered, the second to that of the portion with which it is in contact. Both the symbols v_{12} and v_{13} thus denote the specific volume of the saturated vapour, v_{12} in contact with the liquid, and v_{13} in contact with the solid. Similarly v_{23} and v_{21}, v_{31} and v_{32}, represent the specific volumes of the liquid and of the solid in a state of saturation. Each of these six quantities is a definite function of the temperature alone. The corresponding pressures are

Of vaporization.	Of fusion.	Of sublimation.
$p_{12} = p_{21}$	$p_{23} = p_{32}$	$p_{31} = p_{13}$

These are also functions of the temperature alone. Only at the fundamental temperature do two of these pressures become equal, and therefore equal to the third.

If we represent the relation between these three pressures and the temperature by three curves, the temperatures as abscissæ and the pressures as ordinates, these curves will meet in one point, the *fundamental point*, also called the *triple point*. It is easy to calculate at what angle the curves intersect at the critical point. The inclination of the curves to the abscissa is given by the differential coefficients

$$\frac{dp_{12}}{dT}, \ \frac{dp_{23}}{dT}, \ \frac{dp_{31}}{dT}.$$

We have, therefore, according to equations (111),

$$\frac{dp_{12}}{dT} = \frac{L_{12}}{T(v_1 - v_2)},$$

$$\frac{dp_{23}}{dT} = \frac{L_{23}}{T(v_2 - v_3)},$$

$$\frac{dp_{31}}{dT} = \frac{L_{31}}{T(v_3 - v_1)},$$

where v refers to the fundamental state, and, therefore, requires only one suffix. We can thus find the direction of each curve at the fundamental point if we know the heat of vaporization, of fusion, and of sublimation.

Let us compare, for example, the curve of the vaporization pressure, p_{12}, of water, with its curve of sublimation pressure, p_{13}, near the fundamental point, 0·0075° C. We have, in absolute units,

$L_{12} = 600 \times 4·19 \times 10^7$ (heat of vaporization of water at 0·0075° C.);

$L_{13} = - L_{31} = (80 + 600) \times 4·19 \times 10^7$ (heat of sublimation of ice at 0·0075° C.);

$v_1 = 206000; \ v_2 = 1·00; \ v_3 = 1·09 \ (\S 187); \ T = 273.$

Hence

$$\frac{dp_{12}}{dT} = \frac{600 \times 4·19 \times 10^7 \times 760}{273 \times 206000 \times 1013250} = 0·335,$$

$$\frac{dp_{31}}{dT} = \frac{680 \times 4·19 \times 10^7 \times 760}{273 \times 206000 \times 1013250} = 0·380,$$

in millimeters of mercury. The curve of the sublimation pressure p_{13} is steeper at the fundamental point than the curve of the vaporization pressure p_{12}. For temperatures above the fundamental one, therefore, $p_{13} > p_{12}$; for those below it, $p_{12} > p_{13}$. Their difference is

$$\frac{dp_{13}}{dT} - \frac{dp_{12}}{dT} = \frac{d(p_{13} - p_{12})}{dT} = 0 \cdot 045.$$

If, therefore, the maximum vapour pressure of water be measured above the fundamental point, and of ice below it, the curve of pressure will show an abrupt bend at the fundamental point. This change of direction is measured by the discontinuity of the differential coefficient. At $-1°$ C., $(dT = -1)$, we have, approximately,

$$p_{13} - p_{12} = -0 \cdot 04;$$

i.e. at $-1°$ C. the maximum vapour pressure of ice is about $0 \cdot 04$ mm. less than that of water. This has been verified by experiment.* The existence of a sharp bend in the curve, however, can only be inferred from theory.

§ **189.** We have hitherto extended our investigation only to the different admissible solutions of the equations which express the intrinsic conditions of equilibrium, and have deduced from them the properties of the states of equilibrium to which they lead. We shall now consider the relative merit of these solutions, *i.e.* which of them represents the state of greatest stability. For this purpose we resume our original statement of the problem (§ 165), which is briefly as follows :

Given the total mass M, the total volume V, and the total energy U, it is required to find the state of most stable equilibrium, *i.e.* the state in which the total entropy of the system is an absolute maximum. Instead of V and U, however, it is often more convenient to introduce $v = \dfrac{V}{M}$, the mean specific volume of the system, and $u = \dfrac{U}{M}$, the mean specific energy of the system.

We have found that the conditions of equilibrium admit,

---

* " Wärmetabellen von L. Holborn, K. Scheel, and F. Henning," Braunschweig (Vieweg u. Sohn), 1919, p. 61.

in general, of three kinds of solution, according as the system is split into 1, 2, or 3 states of aggregation. When we come to consider which of these three solutions deserves the preference in a given case, we must remember that the second and third can be interpreted physically only if the values of the masses, as given by the equations (103) and (121), are positive.

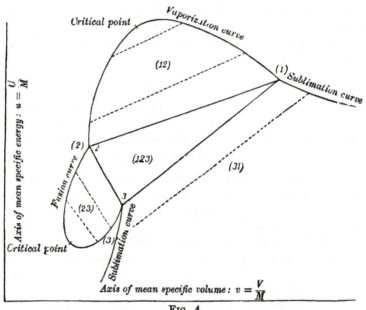

FIG. 4.

This restricts the region of validity of these two solutions. We shall first establish this region of validity, and then prove that within its region the third solution is always preferable to the other two, and, similarly, the second is preferable to the first.

A geometrical representation may facilitate a general survey of the problem. We shall take the mean specific volume, $v = \dfrac{V}{M}$, and the mean specific energy, $u = \dfrac{U}{M}$, of the system as the rectangular coordinate axes. The value of M is here immaterial. Each point of this plane will, then, represent definite values of u and v. Our problem is, therefore, to find the kind of stable equilibrium which will correspond to any given point in this plane.

§ **190.** Let us consider the region of validity of the third solution. The values of the masses given by the equations (121) are

$$M_1 : M_2 : M_3 : M = \begin{vmatrix} 1 & 1 & 1 \\ v & v_2 & v_3 \\ u & u_2 & u_3 \end{vmatrix} : \begin{vmatrix} 1 & 1 & 1 \\ v & v_3 & v_1 \\ u & u_3 & u_1 \end{vmatrix} : \begin{vmatrix} 1 & 1 & 1 \\ v & v_1 & v_2 \\ u & u_1 & u_2 \end{vmatrix} : \begin{vmatrix} 1 & 1 & 1 \\ v_1 & v_2 & v_3 \\ u_1 & u_2 & u_3 \end{vmatrix} \quad (121a)$$

where v_1, v_2, v_3, u_1, u_2, u_3, refer, as hereafter, to the special values which these quantities assume in the fundamental state.

It is obvious from this that the values of M_1, M_2, M_3 can be simultaneously positive only when the point (v, u) lies within the triangle formed by the points (v_1, u_1) (v_2, u_2) and (v_3, u_3). The area of this triangle then represents the region of validity of the third solution. This triangle may be called the *fundamental triangle* of the substance. In Fig. 4 this triangle is represented by (123). The diagram is based on a substance for which, as for water,

$$v_1 > v_3 > v_2 \text{ and } u_1 > u_2 > u_3.$$

§ **191.** We shall now consider the region of validity of the second solution contained in equations (101) and (103). These equations furnish three sets of values for the three possible combinations, and no preference can be given to any one of these. If we consider first the combination of liquid and vapour, the equations referred to become, under our present notation,

$$\left. \begin{array}{l} T_{12} = T_{21} \\ p_{12} = p_{21} \\ \phi_{12} - \phi_{21} = \dfrac{u_{12} - u_{21} + p_{12}(v_{12} - v_{21})}{T_{12}} \end{array} \right\} \quad . \quad (122)$$

$$\left. \begin{array}{l} M_{12} + M_{21} = M \\ M_{12}v_{12} + M_{21}v_{21} = V = Mv \\ M_{12}u_{12} + M_{21}u_{21} = U = Mu \end{array} \right\} \quad . \quad (123)$$

In order to determine the area within which the point (v, u) must lie so that M_{12} and M_{21} may both be positive, we shall find the limits of that area, *i.e.* the curves represented by the conditions $M_{12} = 0$, and $M_{21} = 0$. The introduction of the latter (no liquid mass) gives $M_{12} = M$ and

$$v = v_{12}, u = u_{12} \quad . \quad . \quad . \quad . \quad (124)$$

Since v_{12} and u_{12} are functions of a single variable, the conditions (124) restrict the point (v, u) to a curve, one of the limits of the region of validity. The curve passes through the vertex 1 of the fundamental triangle, because, at the fundamental temperature, $v_{12} = v_1$, and $u_{12} = u_1$. To follow the path of the curve it is necessary to find the differential coefficient $\dfrac{du_{12}}{dv_{12}}$. We have

$$\frac{du_{12}}{dv_{12}} = \left(\frac{\partial u}{\partial v}\right)_{12} + \left(\frac{\partial u}{\partial T}\right)_{12}\frac{dT_{12}}{dv_{12}}.$$

The partial differential coefficients here refer to the independent variables T and v. It follows from (80) and (24) that

$$\frac{du_{12}}{dv_{12}} = T_{12}\left(\frac{\partial p}{\partial T}\right)_{12} - p_{12} + (c_v)_{12}\frac{dT_{12}}{dv_{12}}.$$

By means of this equation the path of the curve (124) may be experimentally plotted by taking T_{12}, or v_{12}, or some other appropriate quantity as independent parameter.

Similarly, the condition $M_{12} = 0$ (no vapour) gives another boundary of the region of validity, viz. the curve,

$$v = v_{21}, u = u_{21},$$

which passes through the vertex 2 of the fundamental triangle, and satisfies the differential equation

$$\frac{du_{21}}{dv_{21}} = T_{12}\left(\frac{\partial p}{\partial T}\right)_{21} - p_{12} + (c_v)_{21}\frac{dT_{12}}{dv_{21}},$$

since $T_{21} = T_{12}$ and $p_{21} = p_{12}$.

The two limiting curves, however, are merely branches of one curve, since they pass into one another at the critical point $(v_{12} = v_{21})$ without forming an angle or cusp at that point, as a further discussion of the values of $\dfrac{du_{12}}{dv_{12}}$ and $\dfrac{du_{21}}{dv_{21}}$, according to § 185, shows. We may, therefore, include the two branches under the name of the *vaporization curve*. Every point (v_{12}, u_{12}) of one branch has a *corresponding point* (v_{21}, u_{21}) on the other, since these two represent the same temperature $T_{12} = T_{21}$, and the same pressure $p_{12} = p_{21}$. This coordination of points on the two branches is given by the equations (122), and has been indicated on our diagram (Fig. 4) by drawing some dotted lines joining corresponding points. In this sense the vertices 1 and 2 of the fundamental triangle are corresponding points, and the critical point is self-corresponding.

This vaporization curve bounds the region of validity of that part of the second solution which refers to liquid in contact with its vapour. Equation (123) makes it obvious that the region of validity lies within the concave side of the curve. The curve has not been produced beyond the vertices 1 and 2 of the fundamental triangle, because we shall see later, that the side 12 of that triangle bounds the area within which this solution gives stable equilibrium. There may be found, quite analogous to the vaporization curve, also a *fusion curve* the two branches of which are represented by

$$v = v_{23}, \; u = u_{23},$$
and
$$v = v_{32}, \; u = u_{32},$$

and a *sublimation curve* represented by

$$v = v_{31}, \; u = u_{31},$$
and
$$v = v_{13}, \; u = u_{13}.$$

The former passes through the vertices 2 and 3, the latter through 3 and 1, of the fundamental triangle. The region of validity of the three parts of the second solution have been marked (12), (23), and (31), respectively, in Fig. 4. The relations which have been specially deduced for the area (12) apply to (23) and (31) as well, only with a corresponding interchange of the suffixes. Some pairs of corresponding points have again been joined by dotted lines. On the fusion curve a critical point has also been marked.

§ **192.** Having thus fixed the region of validity for the second solution, we find that for all points (v, u) outside this region only the first solution admits of physical interpretation. It follows that for such points the first solution represents the stable equilibrium. The areas where such is the case have been marked (1), (2), and (3) in our figure, to signify the gaseous, liquid, and solid states respectively. If there exists a critical point for two states of aggregation, then there is no sharp boundary between them.

§ **193.** We have now to consider the following question : Which of the different states of equilibrium, that may correspond to given values M, v, u (or to a given point of the figure), gives to the system the greatest value of the entropy ? Since each of the three solutions discussed leads to a definite state of the system, we have for each given system (M, v, u) as many values of the entropy as there are solutions applying to it. Denoting these by Φ, Φ', and Φ'', we get for the first solution

$$\Phi = M\phi \quad . \quad . \quad . \quad . \quad (125)$$

for the second :

$$\Phi' = M\phi' = M_{12}\phi_{12} + M_{21}\phi_{21} \quad . \quad . \quad (126)$$

or some other combination of two states of aggregation; for the third :

$$\Phi'' = M\phi'' = M_1\phi_1 + M_2\phi_2 + M_3\phi_3 . \quad . \quad (127)$$

All these quantities are fully determined by the given values of M, v, and u. Now, we can show that for any system (M, v, u) we have $\Phi'' > \Phi' > \Phi$, or $\phi'' > \phi' > \phi$, provided all the partial masses are positive. It is more convenient to deal with the mean specific entropies than with the entropies themselves, because the former, being functions of v and u alone, are quite independent of M.

As a geometrical representation, we may imagine, on the

plane of our figure (Fig. 4), perpendiculars erected at each point (v, u), proportional in length to the values of ϕ, ϕ', and ϕ'' respectively, at that point. The upper ends of these perpendiculars will generate the three surfaces of entropy, ϕ, ϕ', and ϕ''.

§ **194.** We shall show that $\phi' - \phi$ is always positive, *i.e.* that the surface of entropy, ϕ', lies everywhere above the surface ϕ.

While the value of ϕ may be taken directly from (61), which contains the definition of the entropy for homogeneous substances, ϕ' may be found from (126), (122), and (123), in terms of v and u. The surface ϕ' forms three sheets corresponding to the three combinations of two states of aggregation. We shall in the following refer to the combination of vapour and liquid.

With regard to the relative position of the surfaces ϕ and ϕ', it is obvious that they have one curve in common, the projection of which is the vaporization curve. At any point on the vaporization curve we have $v = v_{12}$, $u = u_{12}$, and for the first entropy surface, $\phi = \phi_{12}$; for the second we have, from (123),

$$\mathrm{M}_{21} = 0, \ \mathrm{M}_{12} = \mathrm{M} \ . \quad . \quad . \quad . \quad (128)$$

and, from (126),

$$\phi' = \phi_{12}.$$

In fact, for all points of the vaporization curve, both solutions coincide. The curve of intersection of the surfaces ϕ and ϕ' is represented by

$$v = v_{12}, \ u = u_{12}, \ \phi = \phi_{12},$$

where v, u, and ϕ are the three rectangular coordinates of a point in space. v_{12}, u_{12}, ϕ_{12} depend on a single variable parameter, for example the temperature, $\mathrm{T}_{12} = \mathrm{T}_{21}$. This curve passes through the point (v_1, u_1, ϕ_1), which has the vertex 1 for its projection. A second branch of the same curve is given by the equations

$$v = v_{21}, \ u = u_{21}, \ \phi = \phi_{21},$$

and these branches meet in a point whose projection is the critical point. Each point of one branch has a corresponding point on the other, since both correspond to the same temperature, $T_{12} = T_{21}$, and the same pressure, $p_{12} = p_{21}$. Thus, (v_1, u_1, ϕ_1) and (v_2, u_2, ϕ_2) are corresponding points.

It is further obvious that the surface ϕ' is a ruled surface and is developable. The first may be shown by considering any point with the coordinates

$$v = \frac{\lambda v_{12} + \mu v_{21}}{\lambda + \mu}; \quad u = \frac{\lambda u_{12} + \mu u_{21}}{\lambda + \mu}; \quad \phi = \frac{\lambda \phi_{12} + \mu \phi_{21}}{\lambda + \mu},$$

where λ and μ are arbitrary positive quantities. By giving λ and μ all positive values, we obtain all points of the straight line joining the corresponding points $(v_{12}, u_{12}, \phi_{12})$ and $(v_{21}, u_{21}, \phi_{21})$. But this line lies on the surface ϕ', since all the above values of (v, u, ϕ) satisfy the equations (123) and (126) if we put $M_{12} = \lambda$ and $M_{21} = \mu$. The surface ϕ', then, is formed of the lines joining the corresponding points on the curve in which the surfaces ϕ' and ϕ meet. One of these is the line joining the points (v_1, u_1, ϕ_1) and (v_2, u_2, ϕ_2), the projection of which is the side 12 of the fundamental triangle. At the critical point, the line shrinks to a point, and here the surface ϕ' ends. The other two sheets of the surface are quite similar. One begins at the line joining (v_2, u_2, ϕ_2) and (v_3, u_3, ϕ_3), the other at the line joining (v_3, u_3, ϕ_3) and (v_1, u_1, ϕ_1).

The developability of the surface ϕ' may best be inferred from the following equation of a plane :

$$p_{12}(v - v_{12}) + (u - u_{12}) - T_{12}(\phi - \phi_{12}) = 0,$$

where v, u, ϕ are variable coordinates, while p_{12}, v_{12}, u_{12}, T_{12}, ϕ_{12}, depend, by (122), on one parameter, *e.g.* T_{12}. This plane contains the point $(v_{12}, u_{12}, \phi_{12})$, and by the equations (122) the point $(v_{21}, u_{21}, \phi_{21})$, which are corresponding points, and hence also the line joining them. But it also, by (61), contains the neighbouring corresponding points

$$(v_{12} + dv_{12}, u_{12} + du_{12}, \phi_{12} + d\phi_{12})$$

and

$$(v_{21} + dv_{21}, u_{21} + du_{21}, \phi_{21} + d\phi_{21})$$

hence also the line joining them. Therefore, two consecutive generating lines are coplanar, which is the condition of developability of a surface.

In order to determine the value of $\phi' - \phi$, we shall find the change which this difference undergoes on passing from a point (v, u) to a neighbouring one $(v + \delta v, u + \delta u)$. During this passage we shall keep $M = M_{12} + M_{21}$ constant. This does not affect the generality of the result, since ϕ and ϕ' are functions of v and u only. From (126) we have

$$M\delta\phi' = M_{12}\delta\phi_{12} + M_{21}\delta\phi_{21} + \phi_{12}\delta M_{12} + \phi_{21}\delta M_{21},$$

and, by (61),

$$\delta\phi = \frac{\delta u + p\delta v}{T}.$$

But, by (123),

$$\left.\begin{array}{c} \delta M_{12} + \delta M_{21} = 0 \\ M_{12}\delta v_{12} + M_{21}\delta v_{21} + v_{12}\delta M_{12} + v_{21}\delta M_{21} = M\delta v \\ M_{12}\delta u_{12} + M_{21}\delta u_{21} + u_{12}\delta M_{12} + u_{21}\delta M_{21} = M\delta u \end{array}\right\} \quad (129)$$

Whence, by (122),

$$\delta\phi' = \frac{\delta u + p_{12}\delta v}{T_{12}} \quad . \quad . \quad . \quad . \quad (130)$$

and

$$\delta(\phi' - \phi) = \left(\frac{1}{T_{12}} - \frac{1}{T}\right)\delta u + \left(\frac{p_{12}}{T_{12}} - \frac{p}{T}\right)\delta v. \quad (131)$$

If we now examine the surfaces ϕ and ϕ' in the neighbourhood of their curve of contact, it is evident from the last equation that they touch one another along the whole of this curve. For, at any point of the vaporization curve, we have $v = v_{12}$ and $u = u_{12}$; therefore also

$$T = T_{12}, \text{ and } p = p_{12} \quad . \quad . \quad . \quad (132)$$

and hence, for the entire curve, $\delta(\phi' - \phi) = 0$.

To find the kind of contact between the two surfaces, we form $\delta^2(\phi' - \phi)$ from (131), and apply it to the same points of the curve of contact. In general,

$$\delta^2(\phi' - \phi) = \delta u\left(\frac{\delta T}{T^2} - \frac{\delta T_{12}}{T_{12}{}^2}\right) + \delta v\left(\frac{\delta p_{12}}{T_{12}} - \frac{\delta p}{T} - \frac{p_{12}\delta T_{12}}{T_{12}{}^2} + \frac{p\delta T}{T^2}\right)$$

$$+ \delta^2 u\left(\frac{1}{T_{12}} - \frac{1}{T}\right) + \delta^2 v\left(\frac{p_{12}}{T_{12}} - \frac{p}{T}\right).$$

According to (132) we have, at the points of contact of the surfaces,

$$T^2\delta^2(\phi' - \phi) = \delta u(\delta T - \delta T_{12}) + \delta v(T\delta p_{12} - T\delta p - p\delta T_{12} + p\delta T),$$

or by (61)

$$T\delta^2(\phi' - \phi) = (\delta T - \delta T_{12})\delta\phi + (\delta p_{12} - \delta p)\delta v \quad . \quad (133)$$

All these variations may be expressed in terms of δT and δv, by putting

$$\delta\phi = \frac{c_v}{T}\delta T + \frac{\partial p}{\partial T}\delta v \quad \text{(by 81)},$$

$$\delta p = \frac{\partial p}{\partial T}\delta T + \frac{\partial p}{\partial v}\delta v,$$

$$\delta p_{12} = \frac{dp_{12}}{dT_{12}}\delta T_{12}.$$

We have now to express δT_{12} in terms of δT and δv. Equations (129), simplified by (128), give

$$\frac{\delta u_{12} - \delta u}{u_{12} - u_{21}} = \frac{\delta v_{12} - \delta v}{v_{12} - v_{21}}.$$

In these we put

$$\delta u_{12} = \frac{du_{12}}{dT_{12}}\delta T_{12}, \quad \delta v_{12} = \frac{dv_{12}}{dT_{12}}\delta T_{12} \quad . \quad (134)$$

$$\delta u = c_v\delta T + \frac{\partial u}{\partial v}\delta v,$$

and obtain $\quad \delta T_{12} = \dfrac{c_v\delta T + \left(\dfrac{\partial u}{\partial v} - \dfrac{u_{12} - u_{21}}{v_{12} - v_{21}}\right)\delta v}{\dfrac{du_{12}}{dT_{12}} - \dfrac{u_{12} - u_{21}}{v_{12} - v_{21}}\dfrac{dv_{12}}{dT_{12}}}.$

If we consider that, by (109),

$$\frac{u_{12} - u_{21}}{v_{12} - v_{21}} = T_{12}\frac{dp_{12}}{dT_{12}} - p_{12} \quad . \quad . \quad . \quad (135)$$

$$= T\frac{dp_{12}}{dT_{12}} - p,$$

that, by (80), $\quad \dfrac{\partial u}{\partial v} = T\dfrac{\partial p}{\partial T} - p$

and that
$$\frac{du_{12}}{dT_{12}} = \left(\frac{\partial u}{\partial T}\right)_{12} + \left(\frac{\partial u}{\partial v}\right)_{12} \cdot \frac{dv_{12}}{dT_{12}}$$
$$= c_v + \left(T\frac{\partial p}{\partial T} - p\right)\frac{dv_{12}}{dT_{12}} \quad . \quad . \quad (136)$$

also that
$$\frac{dp_{12}}{dT_{12}} = \frac{\partial p}{\partial T} + \frac{\partial p}{\partial v} \cdot \frac{dv_{12}}{dT_{12}},$$

we obtain
$$\delta T_{12} = \frac{c_v \delta T - T\frac{\partial p}{\partial v} \cdot \frac{dv_{12}}{dT_{12}}\delta v}{c_v - T\frac{\partial p}{\partial v}\left(\frac{dv_{12}}{dT_{12}}\right)^2}.$$

Equation (133), with all variations expressed in terms of δT and δv, finally becomes

$$\delta^2(\phi' - \phi) = -\frac{\partial p}{\partial v} \cdot \frac{c_v}{T} \cdot \frac{\left(\frac{dv_{12}}{dT_{12}}\delta T - \delta v\right)^2}{c_v - T\frac{\partial p}{\partial v}\left(\frac{dv_{12}}{dT_{12}}\right)^2}.$$

This expression is essentially positive, since c_v is positive from its physical meaning, and $\frac{\partial p}{\partial v}$ is negative for any state of equilibrium (§ 169). There is a limiting case, when

$$\frac{dv_{12}}{dT_{12}}\delta T - \delta v = 0,$$

since then,
$$\delta^2(\phi' - \phi) = 0.$$

In this case the variation (δT, δv) obviously takes place along the curve of contact (T_{12}, v_{12}) of the surfaces, and it is evident that then $\phi' = \phi$.

It follows that the surface ϕ', in the vicinity of all points of contact with ϕ, rises above the latter throughout, or that $\phi' - \phi$ is everywhere > 0. This proves that the second solution of the conditions of equilibrium, within its region of validity, *i.e.* in the areas (12), (23), and (31), always represents the stable equilibrium.

§ **195.** Similarly, it may be shown that the third solution, within its region of validity, is preferable to the second one.

The quantities v and u being given, the value of the mean specific entropy, ϕ'', corresponding to this solution is uniquely determined by the equations (127) and (121). The quantities $v_1, v_2, v_3, u_1, u_2, u_3$, and therefore also ϕ_1, ϕ_2, ϕ_3, have definite numerical values, given by equations (120).

In the first place, it is obvious that the surface ϕ'' is the plane triangle formed by the points (v_1, u_1, ϕ_1), (v_2, u_2, ϕ_2), and (v_3, u_3, ϕ_3), the projection of which on the plane of the figure is the fundamental triangle, since any point with the coordinates

$$v = \frac{\lambda v_1 + \mu v_2 + \nu v_3}{\lambda + \mu + \nu},$$

$$u = \frac{\lambda u_1 + \mu u_2 + \nu u_3}{\lambda + \mu + \nu},$$

$$\phi = \frac{\lambda \phi_1 + \mu \phi_2 + \nu \phi_3}{\lambda + \mu + \nu},$$

(λ, μ, ν may have any positive values) satisfies the equations (121) and (127). To show this, we need only put $M_1 = \lambda$, $M_2 = \mu$, $M_3 = \nu$. This plane meets the three sheets of the developable surface ϕ' in the three lines joining the points $(v_1, u_1, \phi_1), (v_2, u_2, \phi_2), (v_3, u_3, \phi_3)$. In fact, by making $\nu = 0$, *i.e.*, by (121), $M_3 = 0$, the third solution coincides with the second; for, then

$$M_1 = M_{12}; \; M_2 = M_{21}; \; v_1 = v_{12}; \; u_1 = u_{12}; \atop v_2 = v_{21}; \; T_1 = T_{12}; \; \text{etc.} \quad\quad (137)$$

If we also put $\mu = 0$, then we have $M_2 = 0$, $v_1 = v$, $u_1 = u$, which means the coincidence of all three surfaces, ϕ'', ϕ', and ϕ.

In order to find the sign of $\phi'' - \phi'$, we again find $\delta(\phi'' - \phi')$ in terms of δu and δv. Equation (127) gives

$$M\delta\phi'' = \phi_1\delta M_1 + \phi_2\delta M_2 + \phi_3\delta M_3 \quad . \quad . \quad (138)$$

where, by (121),

$$\delta M_1 + \delta M_2 + \delta M_3 = 0$$
$$v_1\delta M_1 + v_2\delta M_2 + v_3\delta M_3 = M\delta v$$
$$u_1\delta M_1 + u_2\delta M_2 + u_3\delta M_3 = M\delta u.$$

Multiplying the last of these by $\frac{1}{T_1}$, the second by $\frac{p_1}{T_1}$, and adding to (138), we obtain, with the help of (120),

$$\delta\phi'' = \frac{\delta u + p_1\delta v}{T_1}.$$

This, in combination with (130), gives

$$\delta(\phi'' - \phi') = \left(\frac{1}{T_1} - \frac{1}{T_{12}}\right)\delta u + \left(\frac{p_1}{T_1} - \frac{p_{12}}{T_{12}}\right)\delta v, \quad . \quad (139)$$

if the surface ϕ' is represented by the sheet (12). This equation shows that the surface ϕ'' is a tangent to the sheet (12) along the line joining (v_1, u_1, ϕ_1) and (v_2, u_2, ϕ_2), for all points of this line have $T_1 = T_{12}$, $p_1 = p_{12}$, so that $\delta(\phi'' - \phi')$ vanishes. Thus, we find that the plane ϕ'' is a tangent plane to the three sheets of the surface ϕ'. The curves of contact are the three straight lines which form the sides of the plane triangle ϕ''. We have, from (139), for any point of contact

$$\delta^2(\phi'' - \phi') = \frac{\delta T_{12}}{T_1{}^2}\delta u + \left(\frac{p_1\delta T_{12}}{T_1{}^2} - \frac{\delta p_{12}}{T_1}\right)\delta v$$

since T_1 and p_1 are absolute constants; or

$$T_1{}^2\delta^2(\phi'' - \phi') = \left[\delta u - \left(T_1\frac{dp_{12}}{dT_{12}} - p_1\right)\delta v\right]\delta T_{12}. \quad (140)$$

Now, by the elimination of δM_{12} and δM_{21}, it follows, from (129), that

$$\frac{M_{12}\delta v_{21} + M_{21}\delta v_{21} - M\delta v}{v_{12} - v_{21}} = \frac{M_{12}\delta u_{12} + M_{21}\delta u_{21} - M\delta u}{u_{12} - u_{21}}$$

or, by (135) and (134),

$$M\left[\delta u - \left(T_1\frac{dp_{12}}{dT_{12}} - p_1\right)\delta v\right] = \delta T_{12}\left[M_{12}\frac{du_{12}}{dT_{12}} + M_{21}\frac{du_{21}}{dT_{12}}\right.$$
$$\left. - \left(T_1\frac{dp_{12}}{dT_{12}} - p_1\right)\cdot\left(M_{12}\frac{dv_{12}}{dT_{12}} + M_{21}\frac{dv_{21}}{dT_{12}}\right)\right].$$

Substituting this expression in (140), and replacing $\dfrac{du_{12}}{dT_{12}}$ and $\dfrac{du_{21}}{dT_{21}}$ by their values (136), we obtain

$$\delta^2(\phi'' - \phi') = \frac{\delta T_{12}{}^2}{MT_1{}^2}\Big[M_{12}\big((c_v)_{12} - T_1\big(\tfrac{\partial p}{\partial v}\big)_{12} \cdot \big(\tfrac{dv_{12}}{dT_{12}}\big)^2\big)$$
$$+ M_{21}\big((c_v)_{21} - T_1\big(\tfrac{\partial p}{\partial v}\big)_{21} \cdot \big(\tfrac{dv_{21}}{dT_{12}}\big)^2\big)\Big].$$

This quantity is essentially positive, since M_{12}, M_{21}, as well as c_v, are always positive, and $\dfrac{\partial p}{\partial v}$ always negative for states of equilibrium. There is a limiting case, when $\delta T_{12} = 0$, *i.e.* for a variation along the line of contact of the surfaces ϕ'' and ϕ', as is obvious. It follows that the plane area ϕ'' rises everywhere above the surface ϕ', and that $\phi'' - \phi'$ is never negative. This proves that the third solution within its region of validity (the fundamental triangle of the substance) represents stable equilibrium.

§ **196.** We are now in a position to answer generally the question proposed in § 165 regarding the stability of the equilibrium.

The total mass M, the volume V, and the energy U of a system being given, its corresponding state of stable equilibrium is determined by the position of the point $v = \dfrac{V}{M}$, $u = \dfrac{U}{M}$, in the plane of Fig. 4.

If this point lie within one of the regions (1), (2), or (3), the system behaves as a homogeneous gas, liquid, or solid. If it lie within (12), (23), or (31), the system splits into two different states of aggregation, indicated by the numbers used in the notation of the region. In this case, the common temperature and the ratio of the two heterogeneous portions are completely determined. According to the equation (123), the point (v, u) lies on the straight line joining two corresponding points of the limiting curve. If a straight line be drawn through the given point (v, u), cutting the two branches of that curve in corresponding points, these points give the properties of the two states of aggregation into which the system splits. They have, of course, the same temperature and pressure. The proportion of the two masses, according to the equation (123), is given by the ratio in which the point (v, u) divides the line joining the corresponding points.

If, finally, the point (v, u) lie within the region of the fundamental triangle (123), stable equilibrium is characterized by a division of the system into a solid, a liquid, and a gaseous portion at the fundamental temperature and pressure. The masses of these three portions may then be determined by the equations (121a). It will be seen that their ratio is that of the three triangles, which the point (v, u) makes with the three sides of the fundamental triangle.

The conditions of stable equilibrium of any substance can thus be found, provided its fundamental triangle, its vaporization, fusion, and sublimation curves have been drawn once for all. To obtain a better view of the different relations, isothermal and isobaric curves may be added to the figure. These curves coincide in the regions (12), (23), (31), and form the straight lines joining corresponding points on the limiting curves. On the other hand, the area (123) represents one singular isothermal and isobar (the triple point). In this way we may find that ice cannot exist in stable equilibrium at a higher temperature than the fundamental temperature ($0 \cdot 0075°$ C.), no matter how the pressure may be reduced. Liquid water, on the other hand, may, under suitable pressure, be brought to any temperature without freezing or evaporating.

A question which may also be answered directly is the following. Through what stages will a body pass if subjected to a series of definite external changes? For instance, the behaviour of a body of mass M, when cooled or heated at constant volume V, may be known by observing the line $v = \dfrac{V}{M}$, parallel to the axis of ordinates. The regions which this line traverses show the states through which the body passes, *e.g.* whether the substance melts during the process, or whether it sublimes, etc.*

* Concerning the actual conditions for water, see G. Tammann, *Göttinger Nachrichten*, 1913, p. 99.

CHAPTER III.

SYSTEM OF ANY NUMBER OF INDEPENDENT CONSTITUENTS.

§ **197.** WE proceed to investigate quite generally the equilibrium of a system made up of distinct portions in contact with one another. The system, contrary to that treated of in the preceding chapter, may consist of any number of independent constituents. Following Gibbs, we shall call each one of these portions, inasmuch as it may be considered physically homogeneous (§ 67), a *phase*.* Thus, a quantity of water partly gaseous, partly liquid, and partly solid, forms a system of three phases. The number of phases as well as the states of aggregation is quite arbitrary, although we at once recognize the fact that a system in equilibrium may consist of any number of solid and liquid phases, but only one single *gaseous* phase, for two different gases in contact are never in equilibrium with one another.

§ **198.** A system is characterized by the number of its *independent constituents*, frequently termed *components*, in addition to the number of its phases. The main properties of the state of equilibrium depend upon these. We define the number of independent constituents as follows. First find the number of chemical elements contained in the system, and from these discard, as dependent constituents, all those whose quantity is determined in each phase by the remaining ones. The number of the remaining elements will be the number of independent constituents of the system.† It is immaterial

---

* Concerning the application of the idea of the *phase* to both enantiomorphic forms of optically active substances, see the detailed research by A. Byk, *Zeits. f. phys. Chemie*, **45**, 465, 1903.

† This definition of *independent constituents* of a system satisfies all cases, which deal with true equilibrium, *i.e.* with one which equation (76) satisfies for all thinkable changes.

which of the constituents we regard as independent and which as dependent, since we are here concerned with the number, and not with the kind, of the independent constituents. The question as to the number of the independent constituents has nothing at all to do with the chemical constitution of the substances in the different phases, in particular, with the number of different kinds of molecules.

Thus, a quantity of water in any number of states forms but one independent constituent, however many associations and dissociations of H_2O molecules may occur (it may be a mixture of hydrogen and oxygen or ions), for the mass of the oxygen in each phase is completely determined by that of the hydrogen, and *vice versâ*. As soon as one takes account of the fact that water vapour is at each temperature partly dissociated into hydrogen and oxygen, and that oxygen is more strongly absorbed by the liquid water than is hydrogen, then one obtains, in spite of the fact that only complete H_2O molecules are used to build up the system, different proportions by weight of the elements hydrogen and oxygen in both phases of the system, water and water vapour. There is accordingly not one but two independent constituents in the system. This, of course, is also true if either hydrogen or oxygen is present in excess.

An aqueous solution of sulphuric acid forms a system of three chemical elements, S, H, and O, but contains only two independent constituents, for, in each phase (*e.g.* liquid, vapour, solid), the mass of O depends on that of S and H, while the masses of S and H are not in each phase interdependent. Whether the molecule H_2SO_4 dissociates in any way, or whether hydrates are formed or not, does not change the number of independent constituents of the system.

§ **199.** We denote the number of independent constituents of a system by α. By our definition of this number we see, at once, that each phase of a given system in equilibrium is determined by the masses of each one of its α constituents, the temperature T, and the pressure p. For the sake of

uniformity, we assume that each of the α independent constituents actually occurs in each phase of the system in a
certain quantity, which, in special cases, may become infinitely
small. The selection of the temperature and the pressure
as independent variables, produces a change in the form of
the equations of the last chapter, where the temperature and
the specific volume were considered as the independent
variables. The substitution of the pressure for the volume
is more convenient here, because the pressure is the same for
all phases in free contact, and it can in most cases be more
readily measured.

§ **200.** We shall now consider the thermodynamical
equilibrium of a system, in which the total masses of the α
independent constituents M_1, M_2, . . . M_a are given. Of the
different forms of the condition of equilibrium it is best to use
that expressed by equation (79)

$$\delta\Psi = 0 \quad . \quad . \quad . \quad . \quad (141)$$

which holds, if T and p remain constant, for any change
compatible with the given conditions. The function Ψ is
given in terms of the entropy Φ, the energy U, and the
volume V, by the equation

$$\Psi = \Phi - \frac{U + pV}{T}.$$

§ **201.** Now, let β be the number of phases in the system,
then Φ, U, and V, and therefore also Ψ, are sums of β terms,
each of which refers to a single phase, *i.e.* to a physically
homogeneous body :

$$\Psi = \Psi' + \Psi'' + \ldots + \Psi^\beta . \quad . \quad . \quad (142)$$

where the different phases are distinguished from one another
by dashes. For the first phase,

$$\Psi' = \Phi' - \frac{U' + pV'}{T} \quad . \quad . \quad . \quad (143)$$

Φ', U', V' and Ψ' are completely determined by T, p, and the masses M_1', M_2', ... M_a' of the independent constituents in the phases. As to how they depend on the masses, all we can at present say is, that, if all the masses were increased in the same proportion (say doubled), each of these functions would be increased in the same proportion. Since the nature of the phase remains unchanged, the entropy, the energy, and the volume change in the same proportion as the mass; hence, also, the function Ψ'. In other words Ψ' is a homogeneous function of the masses M_1', M_2', ... M_a' of the first degree but not necessarily linear.*

To express this analytically let us increase all the masses in the same ratio $1 + \varepsilon : 1$, where ε is very small. All changes are then small; and for the corresponding change of Ψ' we obtain

$$\Delta\Psi' = \frac{\partial\Psi'}{\partial M_1}\Delta M_1' + \frac{\partial\Psi'}{\partial M_2}\Delta M_2' + \ldots$$

$$= \frac{\partial\Psi'}{\partial M_1'}\varepsilon M_1' + \frac{\partial\Psi'}{\partial M_2'}\varepsilon M_2' + \ldots$$

But, by supposition, $\Delta\Psi' = \varepsilon\Psi'$,

and, therefore,

$$\Psi' = \frac{\partial\Psi'}{\partial M_1'}M_1' + \frac{\partial\Psi'}{\partial M_2'}M_2' + \ldots + \frac{\partial\Psi'}{\partial M_a'}M_a'. \quad (144)$$

Various forms may be given to this Eulerian equation by further differentiation. The differential coefficients $\frac{\partial\Psi'}{\partial M_1'}$, $\frac{\partial\Psi'}{\partial M_2'}$ ... evidently depend on the constitution of the phase, and not on its total mass, since a change of mass changes both numerator and denominator in the same proportion.

§ **202.** By (142), the condition of equilibrium becomes

$$\delta\Psi' + \delta\Psi'' + \ldots \delta\Psi^\beta = 0 \quad . \quad . \quad (145)$$

or, since the temperature and pressure remain constant,

* $\Psi' = (M_1'^2 + M_2'^2 + \ldots + M_a'^2.)^{\frac{1}{2}}$ is an example of a non-linear homogeneous function of the first degree.

$$\frac{\partial \Psi'}{\partial M_1'}\delta M_1' + \frac{\partial \Psi'}{\partial M_2'}\delta M_2' + \ldots + \frac{\partial \Psi'}{\partial M_a'}\delta M_a'$$

$$+ \frac{\partial \Psi''}{\partial M_1''}\delta M_1'' + \frac{\partial \Psi''}{\partial M_2''}\delta M_2'' + \ldots + \frac{\partial \Psi''}{\partial M_a''}\delta M_a''$$

$$+ \quad . \quad . \quad . \quad . \quad . \quad . \quad . \quad . \quad . \quad . \quad .$$

$$+ \frac{\partial \Psi^\beta}{\partial M_1^\beta}\delta M_1^\beta + \frac{\partial \Psi^\beta}{\partial M_2^\beta}\delta M_2^\beta + \ldots + \frac{\partial \Psi^\beta}{\partial M_a^\beta}\delta M_a^\beta = 0 \quad . \quad (146)$$

If the variation of the masses were quite arbitrary, then the equation could only be satisfied, if all the coefficients of the variations were equal to 0. According to § 200, however, the following conditions exist between them,

$$\left.\begin{aligned} M_1 &= M_1' + M_1'' + \ldots + M_1^\beta \\ M_2 &= M_2' + M_2'' + \ldots + M_2^\beta \\ &\quad . \quad . \quad . \quad . \quad . \quad . \quad . \\ M_a &= M_a' + M_a'' + \ldots + M_a^\beta \end{aligned}\right\} \quad . \quad (147)$$

and, therefore, for any possible change of the system,

$$\left.\begin{aligned} 0 &= \delta M_1' + \delta M_1'' + \ldots + \delta M_1^\beta \\ 0 &= \delta M_2' + \delta M_2'' + \ldots + \delta M_2^\beta \\ &\quad . \quad . \quad . \quad . \quad . \quad . \quad . \\ 0 &= \delta M_a' + \delta M_a'' + \ldots + \delta M_a^\beta \end{aligned}\right\} \quad . \quad (148)$$

For the expression (146) to vanish, the necessary and sufficient condition is

$$\left.\begin{aligned} \frac{\partial \Psi'}{\partial M_1'} &= \frac{\partial \Psi''}{\partial M_1''} = \ldots = \frac{\partial \Psi^\beta}{\partial M_1^\beta} \\ \frac{\partial \Psi'}{\partial M_2'} &= \frac{\partial \Psi''}{\partial M_2''} = \ldots = \frac{\partial \Psi^\beta}{\partial M_2^\beta} \\ &\quad . \quad . \quad . \quad . \quad . \quad . \\ \frac{\partial \Psi'}{\partial M_a'} &= \frac{\partial \Psi''}{\partial M_a''} = \ldots = \frac{\partial \Psi^\beta}{\partial M_a^\beta} \end{aligned}\right\} \quad . \quad (149)$$

There are for each independent constituent $(\beta - 1)$ equations which must be satisfied, and therefore for all the α independent constituents $\alpha(\beta - 1)$ conditions. Each of these equations refers to the transition from one phase into another, and asserts that this particular transition does not take place in nature. This condition depends, as it must, on the internal constitution of the phase, and not on its total mass.

Since the equations in a single row with regard to a particular constituent may be arranged in any order, it follows that, if a phase be in equilibrium as regards a given constituent with two others, these two other phases are in equilibrium with one another with regard to that constituent (they coexist). This shows that, since any system in equilibrium can have only one gaseous phase, two coexisting phases, *e.g.* two liquids which form two layers, must emit the same vapour. For, since each phase is in equilibrium with the other, and also with its own vapour with respect to all constituents, it must also coexist with the vapour of the second phase. The coexistence of solid and liquid phases may, therefore, be settled by comparing their vapours.

§ 203. It is now easy to see how the state of equilibrium of the system is determined, in general, by the given external conditions (147), and the conditions of equilibrium (149). There are α of the former and $\alpha(\beta - 1)$ of the latter, a total of $\alpha\beta$ equations. On the other hand, the state of the β phases depends on $(\alpha\beta + 2)$ variables, viz. on the $\alpha\beta$ masses. $M_1', \ldots M_a^\beta$, the temperature T, and the pressure p. After all conditions have been satisfied, two variables still remain undetermined. In general, the temperature and the pressure may be arbitrarily chosen, but in special cases, as will be shown presently, these are no longer arbitrary, and in such cases two other variables, as the total energy and the total volume of the system, are undetermined. By disposing of the values of the arbitrary variables we completely determine the state of the equilibrium.

§ 204. The $\alpha\beta + 2$ variables, which control the state of the system, may be separated into those which merely govern the composition of the phases (*internal* variables), and those which determine only the total masses of the phases (*external* variables). The number of the former is $(\alpha - 1)\beta + 2$, for in each of the β phases there are $\alpha - 1$ ratios between its α

independent constituents, to which must be added temperature and pressure. The number of the external variables is β, viz. the total masses of all the phases. Altogether :

$$(\alpha - 1)\beta + 2 + \beta = \alpha\beta + 2.$$

We found that the $\alpha(\beta - 1)$ equations (149) contain only internal variables, and, therefore, after these have been satisfied, there remain

$$[(\alpha - 1)\beta + 2] - [\alpha(\beta - 1)] = \alpha - \beta + 2$$

of the internal variables, undetermined. This number cannot be negative, for otherwise the number of the internal variables of the system would not be sufficient for the solution of the equations (149). It, therefore, follows that

$$\beta \leqq \alpha + 2.$$

The number of the phases, therefore, cannot exceed the number of the independent constituents by more than two; or, a system of α independent constituents will contain at most $(\alpha + 2)$ phases. In the limiting case, where $\beta = \alpha + 2$, the number of the internal variables are just sufficient to satisfy the internal conditions of equilibrium (149). Their values in the state of equilibrium are completely determined quite independently of the given external conditions. Decreasing the number of phases by one increases the number of the indeterminate internal variables by one.

This proposition, first propounded by Gibbs and universally known as the *phase rule*, has been amply verified, especially by the experiments of Bakhuis Roozeboom.*

§ 205. We shall consider, first, the limiting case :

$$\beta = \alpha + 2$$

(Non-variant systems). Since all the internal variables are completely determined, they form an $(\alpha + 2)$-*ple point*. Change of the external conditions, as heating, compression,

---

* See the book, " Die heterogenen Gleichgewichte vom Standpunkte der Phasenlehre," by Bakhuis Roozeboom, Braunschweig, Vieweg & Sohn, 1904.

further additions of the substances, alter the total masses of the phases, but not their internal nature, including temperature and pressure. This holds until the mass of some one phase becomes zero, and therewith completely vanishes from the system.

If $\alpha = 1$, then $\beta = 3$. A single constituent may split into three phases at most, forming a triple point. An example of this is a substance existing in the three states of aggregation, all in contact with one another. For water it was shown in § 187, that at the triple point the temperature is 0·0075° C., and the pressure 4·58 mm. of mercury. The three phases need not, however, be different states of aggregation. Sulphur, for instance, forms several modifications in the solid state. Each modification constitutes a separate phase, and the proposition holds that two modifications of a substance can coexist with a third phase of the same substance, for example, its vapour, only at a definite temperature and pressure.

A quadruple point is obtained when $\alpha = 2$. Thus, the two independent constituents, SO_2 (sulphur dioxide) and H_2O, form the four coexisting phases : $SO_2,7H_2O$ (solid) SO_2 dissolved in H_2O (liquid), SO_2 (liquid), SO_2 (gaseous), at a temperature of 12·1° C. and a pressure of 1773 mm. of mercury. The question as to the formation of hydrates by SO_2 in aqueous solution does not influence the application of the phase rule (see § 198). Another example is an aqueous solution of sodium chloride in contact with solid salt, ice and water vapour.

Three independent constituents ($\alpha = 3$) lead to a quintuple point. Thus Na_2SO_4, $MgSO_4$, and H_2O give the double salt $Na_2Mg(SO_4)_2 4H_2O$ (astrakanite), the crystals of the two simple salts, aqueous solution, and water vapour, at a temperature of 21·5° C. and a pressure of about 10 mm. of mercury.

§ 206. We shall now take the case

$$\beta = \dot{\alpha} + 1,$$

that is, α independent constituents form $\alpha + 1$ phases (Univariant systems). The composition of all the phases is then completely determined by a single variable, *e.g.* the temperature or the pressure. This case is generally called *perfect heterogeneous* equilibrium.

If $\alpha = 1$, then $\beta = 2$: one independent constituent in two phases, *e.g.* a liquid and its vapour. The pressure and the density of the liquid and of the vapour depend on the temperature alone, as was pointed out in the last chapter. Evaporation involving chemical decomposition also belongs to this class, since the system contains only one independent constituent. The evaporation of solid NH_4Cl is a case in point. Unless there be present an excess of hydrochloric acid or ammonia gas, there will be for each temperature a quite definite dissociation pressure.

If $\alpha = 2$, then $\beta = 3$, for instance when the solution of a salt is in contact with its vapour and with the solid salt, or when two liquids that cannot be mixed in all proportions (ether and water) are in contact with their common vapour. Vapour pressure, density and concentration in each phase, are here functions of the temperature alone.

§ 207. We often take the pressure instead of the temperature as the variable which controls the phases in perfect heterogeneous equilibrium; namely, in systems which do not possess a gaseous phase, so-called *condensed* systems. Upon these the influence of the pressure is so slight that, under ordinary circumstances, it may be considered as given, and equal to that of the atmosphere. The phase rule, therefore, gives rise to the following proposition : *A condensed system of α independent constituents forms $\alpha + 1$ phases at most, and is then completely determined, temperature included.* The melting point of a substance, and the point of transition from one allotropic modification to another, are examples of $\alpha = 1$, $\beta = 2$. The point at which the cryohydrate (ice and solid salt) separates out from the solution of a salt, and also the point at which two liquid layers in contact begin to precipitate a solid (*e.g.* $AsBr_3$, and H_2O) are examples of $\alpha = 2$, $\beta = 3$.

We have an example of $\alpha = 3$, $\beta = 4$ when a solution of two salts, capable of forming a double salt, is in contact with the solid simple salts, and also with the double salt.

§ 208. If

$$\beta = \alpha,$$

then α independent constituents form α phases (Divariant systems). The internal nature of all the phases depends on two variables, *e.g.* on temperature and pressure. Any homogeneous substance furnishes an example of $\alpha = 1$. A liquid solution of a salt in contact with its vapour is an example of $\alpha = 2$. The temperature and the pressure determine the concentration in the vapour as well as in the liquid. The concentration of the liquid and either the temperature or the pressure are frequently chosen as the independent variables. In the first case, we say that a solution of given concentration and given temperature emits a vapour of definite composition and definite pressure; and in the second case, that a solution of given concentration and given pressure has a definite boiling point, and at this temperature a vapour of definite composition may be distilled off.

Corresponding regularities hold when the second phase is not gaseous but solid or liquid, as in the case of two liquids which do not mix in all proportions. The internal nature of the two phases, in our example the concentrations in the two layers of the liquids, depends on two variables—pressure and temperature. If, under special circumstances, the concentrations become equal, a phenomenon is obtained which is quite analogous to that of the critical point of a homogenous substance (critical solution temperature of two liquids).

§ 209. Let us now consider briefly the case

$$\beta = \alpha - 1,$$

where the number of phases is one less than the number of the independent constituents, and the internal nature of all phases depends on a third arbitrary variable, besides temperature and pressure. Thus, $\alpha = 3$, $\beta = 2$ for an aqueous solution of two isomorphous substances (potassium chlorate and thallium chlorate) in contact with a mixed crystal. The concentration of the solution under atmospheric pressure and at a given temperature will vary according to the composition of the mixed crystal. We cannot, therefore, speak of a saturated solution of the two substances of definite composition. However, should a second solid phase—for instance, a mixed crystal of different composition—separate out, the internal nature of the system will be determined by temperature and pressure alone. The experimental investigation of the equilibrium of such systems may enable us to decide whether a precipitate from a solution of two salts form one phase—for example, a mixed crystal of changing concentration—or whether the two substances are to be considered as two distinct phases in contact. If, at a given temperature and pressure, the concentration of the liquid in contact were quite definite, it would represent the former case, and, if not, the latter.

§ **210.** If the expressions for the functions Ψ', Ψ'', ... for each phase were known, the equations (149) would give every detail regarding the state of the equilibrium. This, however, is by no means the case, for, regarding the relations between these functions and the masses of the constituents in the individual phases, all we can, in general, assert is that they are homogeneous functions of the first degree (§ 201). We can, however, tell, as we have seen in § 152a, equation (79b), how they depend upon temperature and pressure, since their differential coefficients with respect to T and p can be given. This point leads to far-reaching conclusions concerning the variation of the equilibrium with temperature and pressure.

If the problem is to express the characteristic function Ψ, with the help of physical measurements, as a function of all the independent variables, it is best to start from equation (75), which gives the relation between Ψ, the entropy Φ and Gibbs' heat function $H = U + pV$ (§ 100) :

$$\Psi = \Phi - \frac{H}{T}. \qquad . \quad . \quad . \quad . \quad (150)$$

H in this equation, as well as Φ, can be determined by heat measurements. If C_p denote the heat capacity at constant pressure, then by (26) :

$$C_p = \left(\frac{\partial H}{\partial T}\right)_p \quad \cdot \quad \cdot \quad \cdot \quad \cdot \quad (150a)$$

Also by (84a) :

$$C_p = T\left(\frac{\partial \Phi}{\partial T}\right)_p \quad \cdot \quad \cdot \quad \cdot \quad \cdot \quad (150b)$$

In place of (150) we may write

$$\Psi = \int \frac{C_p}{T} dT - \frac{1}{T}\int C_p dT \quad \cdot \quad \cdot \quad \cdot \quad (150c)$$

Both integrations are to be performed under constant pressure. The finding of the characteristic function Ψ, and therewith all thermodynamical properties of the system considered, is here based upon the measurement of the heat capacity, C_p, of the system for all values of T and p. The additive constant of integration has still to be considered. This could depend on p, and, besides, on the chemical composition of the system. The dependance on p is given by the first equation of (79b) by a measurement of the volume V. The dependance on the chemical composition can be concluded from the measurement of such processes as are accompanied by chemical changes of state.

Finally, there remains in the expression for Ψ still an additive term of the form $a + \dfrac{b}{T}$, which is quite arbitrary (§ 152).

§ 211. These relations may be used to determine how the equilibrium depends on the temperature and pressure. For this purpose we shall distinguish between two different kinds of infinitely small changes. The notation δ will refer, as hitherto, to a change of the masses M_1', M_2', . . . M_a^β. consistent with the given external conditions, and, therefore, consistent with the equations (148), temperature and pressure being kept constant, *i.e.* $\delta T = 0$ and $\delta p = 0$. The state, to which this variation leads, need not be one of equilibrium, and the equations (149) need not, therefore, apply to it. The notation d, on the other hand, will refer to a change from one state of equilibrium to another, only slightly different from it. All external conditions, including temperature and pressure, may be changed in any arbitrary manner.

The problem is now to find the conditions of equilibrium of this second state, and to compare them with those of the original state. Since the condition of equilibrium of the first state is

$$\delta \Psi = 0,$$

the condition for the second state is

$$\delta(\Psi + d\Psi) = 0,$$

hence
$$\delta d\Psi = 0 \quad . \quad . \quad . \quad . \quad (151)$$

But

$$d\Psi = \frac{\partial \Psi}{\partial T} dT + \frac{\partial \Psi}{\partial p} dp + \sum^{\beta} \frac{\partial \Psi'}{\partial M_1'} dM_1' + \frac{\partial \Psi'}{\partial M_2'} dM_1' + \cdots$$

where $\sum$ denotes the summation over all the β phases of the system, while the summation over the α constituents of a single phase is written out at length. This becomes, by (79*b*),

$$d\Psi = \frac{U + pV}{T^2} dT - \frac{V}{T} dp + \sum^{\beta} \frac{\partial \Psi'}{\partial M_1'} dM_1' + \frac{\partial \Psi'}{\partial M_2'} dM_2' + \cdots$$

The condition of equilibrium (151) therefore becomes

$$\frac{\delta U + p\delta V}{T^2} dT - \frac{\delta V}{T} dp + \sum^{\beta} dM_1' \delta \frac{\partial \Psi'}{\partial M_1'}$$
$$+ dM_2' \delta \frac{\partial \Psi'}{\partial M_2'} + \cdots = 0 \ (152)$$

All variations of dT, dp, dM_1', dM_2', $\ldots$ disappear because $\delta T = 0$ and $\delta p = 0$, and because in the sum

$$\frac{\partial \Psi'}{\partial M_1'} \delta dM_1' + \frac{\partial \Psi'}{\partial M_2'} \delta dM_2' + \cdots$$

$$+ \frac{\partial \Psi''}{\partial M_1''} \delta dM_1'' + \frac{\partial \Psi''}{\partial M_2''} \delta dM_2'' + \cdots$$

$$+ \quad \cdots \cdots \cdots \cdots \cdots$$

$$+ \frac{\partial \Psi^\beta}{\partial M_1^\beta} \delta dM_1^\beta + \frac{\partial \Psi^\beta}{\partial M_2^\beta} \delta dM_2^\beta + \cdots$$

each vertical column vanishes. Taking the first column for example, we have, by (149),

$$\frac{\partial \Psi'}{\partial M_1'} = \frac{\partial \Psi''}{\partial M_1''} = \cdots = \frac{\partial \Psi^\beta}{\partial M_1^\beta},$$

and also, by (148),

$$\delta dM_1' + \delta dM_1'' + \cdots + \delta dM_1^\beta$$
$$= d(\delta M_1' + \delta M_1'' + \cdots + \delta M_1^\beta) = 0.$$

Furthermore, since, by the first law, $\delta U + p \delta V$ represents Q, the heat absorbed by the system during the virtual change, the equation (152) may also be written

$$\frac{Q}{T^2} dT - \frac{\delta V}{T} dp + \sum^\beta dM_1' \delta \frac{\partial \Psi'}{\partial M_1'} + dM_2' \delta \frac{\partial \Psi'}{\partial M_2'} + \cdots = 0. \quad (153)$$

This equation shows how the equilibrium depends on the temperature, and the pressure, and on the masses of the independent constituents of the system. It shows, in the first place, that the influence of the temperature depends essentially on the heat of reaction which accompanies a virtual change of state. If this be zero, the first term vanishes, and a change of temperature does not disturb the equilibrium. If Q change sign, the influence of the temperature is also reversed. It is quite similar with regard to the influence of the pressure, which, in its turn, depends essentially on the change of volume, δV, produced by a virtual isothermal and isobaric change of state.

§ 212. We shall now apply the equation (153) to several

special cases; first, to those of perfect heterogeneous equilibrium, which are characterized (§ 206) by the relation

$$\beta = \alpha + 1.$$

The internal nature of all the phases, including the pressure, is determined by the temperature alone. An isothermal, infinitely slow compression, therefore, changes only the total masses of the phases, but does not change either the composition or the pressure. We shall choose a change of this kind as the virtual change of state. In this special case it leads to a new state of equilibrium. The internal nature of all the phases, as well as the temperature and pressure, remain constant, and therefore the variations of the functions $\frac{\partial \Psi'}{\partial M_1''}, \frac{\partial \Psi'}{\partial M_2''}, \ldots$ are all equal to zero, since these quantities depend only on the nature of the phases. The equation (153) therefore becomes

$$\frac{dp}{dT} = \frac{Q}{T\delta V} \quad \cdots \quad \cdots \quad (154)$$

This means that the heat of reaction in a variation that leaves the composition of all phases unchanged, divided by the change of volume of the system and by the absolute temperature, gives the rate of change of the equilibrium pressure with the temperature. Where application of heat increases the volume, as in the case of evaporation, the equilibrium pressure increases with temperature; in the opposite case, as in the melting of ice, it decreases with increase of temperature.

§ 213. In the case of one independent constituent ($\alpha = 1$, and $\beta = 2$), equation (154) leads immediately to the laws discussed at length in the preceding chapter; namely, those concerning the heat of vaporization, of fusion, and of sublimation. If, for instance, the liquid form the first phase, the vapour the second phase, and L denote the heat of vaporization per unit mass, we have

$$Q = L\delta M''$$
$$\delta V = (v'' - v')\delta M''$$

where v' and v'' are the specific volumes of liquid and vapour, and $\delta M''$ the mass of vapour formed during the isothermal and isobaric change of state. Hence, by (154),

$$L = T\frac{dp}{dT}(v'' - v'),$$

which is identical with the equation (111).

This, of course, applies to chemical changes as well, whenever the system under consideration contains one constituent in two distinct phases; for example, to the vaporization of ammonium chloride (first investigated with regard to this law by Horstmann), which decomposes into hydrochloric acid and ammonia; or to the vaporization of ammonium carbamate, which decomposes into ammonia and carbon dioxide. Here L of our last equation denotes the heat of dissociation, and p the dissociation pressure, which depends only on the temperature.

§ **214.** We shall also consider the perfect heterogeneous equilibrium of two independent constituents ($\alpha = 2$, $\beta = 3$); for example, water (suffix 1) and a salt (suffix 2) in three phases; the first, an aqueous solution (M_1' the mass of the water, M_2' that of the salt); the second, water vapour (mass M_1''); the third, solid salt (mass M_2'''). For a virtual change, therefore,

$$\delta M_1' + \delta M_1'' = 0, \text{ and } \delta M_2' + \delta M_2''' = 0.$$

According to the phase rule, the concentration of the solution $\left(\frac{M_2'}{M_1'} = c\right)$, as well as the vapour pressure (p), is a function of the temperature alone. By (154), the heat absorbed (T, p, c remaining constant) is

$$Q = T \cdot \frac{dp}{dT} \cdot \delta V \quad . \quad . \quad . \quad . \quad (155)$$

Let the virtual change consist in the evaporation of a small quantity of water,

$$\delta M_1'' = -\delta M_1'.$$

Then, since the concentration also remains constant, the quantity of salt

$$\delta M_2''' = - \delta M_2' = - c\delta M_1' = c\delta M_1''$$

is precipitated from the solution. All variations of mass have here been expressed in terms of $\delta M_1''$.

The total volume of the system

$$V = v'(M_1' + M_2') + v''M_1'' + v'''M_2''',$$

where v', v'', and v''' are the specific volumes of the phases, is increased by

$$\delta V = v'(\delta M_1' + \delta M_2') + v''\delta M_1'' + v'''\delta M_2'''$$
$$\delta V = [(v'' + cv''') - (1 + c)v']\delta M_1'' \quad . \quad . \quad . \quad (156)$$

If L be the quantity of heat that must be applied to evaporate unit mass of water from the solution, and to precipitate the corresponding quantity of salt, under constant pressure, temperature, and concentration, then the equation (155), since

$$Q = L\delta M_1'',$$

becomes $\qquad L = T\dfrac{dp}{dT}(v'' + cv''' - (1 + c)v').$

A useful approximation is obtained by neglecting v' and v''', the specific volumes of the liquid and solid, in comparison with v'', that of the vapour, and considering the latter as a perfect gas. By (14),

$$v'' = \frac{R}{m} \cdot \frac{T}{p}$$

(R = gas constant, m = the molecular weight of the vapour), and we obtain

$$L = \frac{R}{m}T^2 \cdot \frac{d\log p}{dT} \quad . \quad . \quad . \quad . \quad (157)$$

§ **215.** Conversely, L is at the same time the quantity of heat given out when unit mass of water vapour combines, at constant temperature and pressure, with the quantity of salt necessary to form a saturated solution.

This process may be accomplished directly, or in two steps, viz. by condensing unit mass of water vapour into pure water, and then dissolving the salt in the water. According to the first law of thermodynamics, since the initial and final states are the same in both cases, the sum of the heat given out and the work done is the same.

In the first case the heat given out is L, the work done, $-p\dfrac{\delta V}{\delta M_1''}$; and the sum of these, by the approximation used above, is

$$\frac{R}{m}T^2 . \frac{d \log p}{dT} - pv'' \quad . \quad . \quad . \quad . \quad (158)$$

To calculate the same sum for the second case, we must in the first place note that the vapour pressure of a solution is different from the vapour pressure of pure water at the same temperature. It will, in fact, in no case be greater, but smaller, otherwise the vapour would be supersaturated. Denoting the vapour pressure of pure water at the temperature T by p_0, then $p < p_0$.

We shall now bring, by isothermic compression, unit mass of water vapour from pressure p and specific volume v'' to pressure p_0 and specific volume v_0'', i.e. to a state of saturation. Work is thereby done on the substance, and heat is given out. The sum of both, which gives the decrease of the energy of the vapour, is zero, if we again assume that the vapour behaves as a perfect gas, i.e. that its energy remains constant at constant temperature. If we then condense the water vapour of volume v_0'', at constant temperature T and constant pressure p_0, into pure water, the sum of the heat given out and work spent at this step is, by equation (112),

$$\frac{R}{m}T^2 . \frac{d \log p_0}{dT} - p_0 v_0'' . \quad . \quad . \quad . \quad (159)$$

No appreciable external effects accompany the further change of the liquid water from pressure p_0 to pressure p.

If, finally, we dissolve salt sufficient for saturation in the newly formed unit of water, at constant temperature T and constant pressure p, the sum of the heat and work is simply the heat of solution

$$\lambda \quad . \quad . \quad . \quad . \quad . \quad . \quad (160)$$

By the first law, the sum of (159) and (160) must be equal to (158), hence

$$\frac{R}{m}T^2 . \frac{d \log p_0}{dT} - p_0 v_0{}'' + \lambda = \frac{R}{m}T^2 . \frac{d \log p}{dT} - pv'' ;$$

or, if we apply Boyle's law $p_0 v_0{}'' = pv''$,

$$\lambda = \frac{R}{m}T^2 \frac{d \log \dfrac{p}{p_0}}{dT} . \quad . \quad . \quad . \quad (161)$$

This formula, first established by Kirchhoff, gives the heat evolved when salt sufficient for saturation is dissolved in 1 gr. of pure water.

To express λ in calories, R must be divided by the mechanical equivalent of heat, J. By (34), $\dfrac{R}{J} = 1 \cdot 985$, and since $m = 18$, we have

$$\lambda = 0 \cdot 11 T^2 . \frac{d \log \dfrac{p}{p_0}}{dT} \text{ cal.}$$

It is further worthy of notice that p, the vapour pressure of a saturated solution, is a function of the temperature alone, since c, the concentration of a saturated solution, changes in a definite manner with the temperature.

The quantities neglected in this approximation may, if necessary, be put in without any difficulty.

§ **216.** We proceed now to the important case of two independent constituents in two phases ($\alpha = 2$, $\beta = 2$). We assume, for the present, that both constituents are contained in both phases in appreciable quantity, having the masses $M_1{}'$, $M_2{}'$ in the first; $M_1{}''$, $M_2{}''$, in the second phase. The internal variables are the temperature, the pressure, and the concentrations of the second constituent in both phases;

$$c' = \frac{M_2'}{M_1'}, \text{ and } c'' = \frac{M_2''}{M_1''} \quad . \quad . \quad . \quad (162)$$

According to the phase rule, two of the variables, T, p, c', c'', are arbitrary. The others are thereby determined.

Equation (153) leads to the following law regarding the shift of the equilibrium corresponding to any change of the external conditions :

$$\frac{Q}{T^2}dT - \frac{\delta V}{T}dp + dM_1'\delta\frac{\partial\Psi'}{\partial M_1'} + dM_2'\delta\frac{\partial\Psi'}{\partial M_2'} + dM_1''\delta\frac{\partial\Psi''}{\partial M_1''}$$
$$+ dM_2''\delta . \frac{\partial\Psi''}{\partial M_2''} = 0 . \quad . \quad (163)$$

Here, for the first phase,

$$\left.\begin{array}{l} \delta\dfrac{\partial\Psi'}{\partial M_1'} = \dfrac{\partial^2\Psi'}{\partial M_1'^2}\delta M_1' + \dfrac{\partial^2\Psi'}{\partial M_1'\partial M_2'}\delta M_2' \\[2ex] \delta\dfrac{\partial\Psi'}{\partial M_2'} = \dfrac{\partial^2\Psi'}{\partial M_1'\partial M_2'}\delta M_1' + \dfrac{\partial^2\Psi'}{\partial M_2'^2}\delta M_2' \end{array}\right\} \quad . \quad (164)$$

Certain simple relations hold between the derived functions of Ψ' with respect to M_1' and M_2'. For, since, by (144),

$$\Psi' = M_1'\frac{\partial\Psi'}{\partial M_1'} + M_2'\frac{\partial\Psi'}{\partial M_2'},$$

partial differentiation with respect to M_1' and M_2' gives

$$0 = M_1'\frac{\partial^2\Psi'}{\partial M_1'^2} + M_2'\frac{\partial^2\Psi'}{\partial M_1'\partial M_2'},$$
$$0 = M_1'\frac{\partial^2\Psi'}{\partial M_1'\partial M_2'} + M_2'\frac{\partial^2\Psi'}{\partial M_2'^2},$$

If we put, for shortness,

$$M_1'\frac{\partial^2\Psi'}{\partial M_1'\partial M_2'} = \psi' \quad . \quad . \quad . \quad (165)$$

a quantity depending only on the nature of the first phase,

on T, p, and c', and not on the masses M_1' and M_2' individually,* we have

$$\left.\begin{array}{l} \dfrac{\partial^2 \Psi'}{\partial M_1' \partial M_2'} = \dfrac{\psi'}{M_1'} \\[2ex] \dfrac{\partial^2 \Psi'}{\partial M_1'^2} = -\dfrac{M_2'}{M_1'^2} \cdot \psi' \\[2ex] \dfrac{\partial^2 \Psi'}{\partial M_2'^2} = -\dfrac{\psi'}{M_2'} \end{array}\right\} \quad \cdot \quad \cdot \quad \cdot \quad (166)$$

Analogous equations hold for the second phase if we put

$$\psi'' = M_1'' \cdot \frac{\partial^2 \Psi''}{\partial M_1'' \partial M_2''}.$$

§ 217. With respect to the quantities ψ' and ψ'' all we can immediately settle is their sign. According to § 147, Ψ is a maximum in stable equilibrium if only processes at constant temperature and constant pressure be considered. Hence

$$\delta^2 \Psi < 0 \quad \cdot \quad \cdot \quad \cdot \quad \cdot \quad (167)$$

But

$$\Psi = \Psi' + \Psi'',$$

whence

$$\delta\Psi = \frac{\partial \Psi'}{\partial M_1'} \delta M_1' + \frac{\partial \Psi'}{\partial M_2'} \delta M_2' + \frac{\partial \Psi''}{\partial M_1''} \delta M_1'' + \frac{\partial \Psi''}{\partial M_2''} \delta M_2''$$

and

$$\delta^2\Psi = \frac{\partial^2 \Psi'}{\partial M_1'^2}\delta M_1'^2 + 2\frac{\partial^2 \Psi'}{\partial M_1' \partial M_2'}\delta M_1' \delta M_2' + \frac{\partial^2 \Psi'}{\partial M_2'^2}\delta M_2'^2$$
$$+ \frac{\partial^2 \Psi''}{\partial M_1''^2}\delta M_1''^2 + 2\frac{\partial^2 \Psi''}{\partial M_1'' \partial M_2''}\delta M_1'' \delta M_2'' + \frac{\partial^2 \Psi''}{\partial M_2''^2}\delta M_2''^2.$$

If we introduce the quantities ψ' and ψ'', then

$$\delta^2\Psi = -M_2'\psi'\left(\frac{\delta M_1'}{M_1'} - \frac{\delta M_2'}{M_2'}\right)^2 - M_2''\psi''\left(\frac{\delta M_1''}{M_1''} - \frac{\delta M_2''}{M_2''}\right)^2.$$

This relation shows that the inequality (167) is satisfied, and only then, if both ψ' and ψ'' are positive.

* The general integral of $\Psi' = M_1' \dfrac{\partial \Psi'}{\partial M_1'} + M_2' \dfrac{\partial \Psi'}{\partial M_2'}$, is $\Psi' = M_2' f\left(\dfrac{M_1'}{M_2'}\right)$.—Tr.

§ **218.** There are on the whole two kinds of changes possible, according as the first or the second constituent passes from the first to the second phase. We have, for the first,

$$\delta M_1' = - \delta M_1''; \quad \delta M_2' = \delta M_2'' = 0; \quad \cdot \quad \cdot \quad (168)$$

and for the second,

$$\delta M_1' = \delta M_1'' = 0; \quad \delta M_2' = - \delta M_2''.$$

We shall distinguish Q, the heat absorbed, and δV, the change of volume, in these two cases by the suffixes 1 and 2. In the first case, the law for the displacement of the equilibrium by (163), (164), (168), (166), and (162), reduces to

$$\frac{Q_1}{T^2}dT - \frac{\delta_1 V}{T}dp - \delta M_1''(\psi'dc' - \psi''dc'') = 0$$

and, introducing for shortness the finite quantities

$$L_1 = \frac{Q_1}{\delta M_1'''}, \quad v_1 = \frac{\delta_1 V}{\delta M_1'''}, \quad \cdot \quad \cdot \quad (169)$$

i.e. the ratios of the heat absorbed and of the change of volume to the mass of the first constituent, which passes from the first to the second phase, we have

$$\frac{L_1}{T^2}dT - \frac{v_1}{T}dp - \psi'dc' + \psi''dc'' = 0 \quad \cdot \quad \cdot \quad (170)$$

Similarly for the second constituent passing into the second phase, we get

$$\frac{L_2}{T^2}dT - \frac{v_2}{T}dp + \psi'\frac{dc'}{c'} - \psi''\frac{dc''}{c''} = 0 \quad \cdot \quad \cdot \quad (171)$$

These are the two relations, according to the phase rule, which connect the four differentials dT, dp, dc', dc'' in any displacement of the equilibrium.

§ **219.** To show the application of these laws, let us consider a mixture of two constituents (*e.g.* water and alcohol), in two phases, the first a liquid, the second a vapour. According to the phase rule, two of the four variables T, p, c', c'' are determined by the other two.

If the liquid mixture is evaporated, under constant pressure, infinitely slowly by the necessary addition of heat, then at each temperature the concentrations in both phases are completely determined. By the continuous application of heat and the consequent rise of temperature ($dT > 0$) the concentrations c' and c'' change in the following way

$$\frac{dc'}{c'} = \frac{L_1 + c''L_2}{c' - c''} \frac{dT}{T^2\psi'} \quad \cdot \quad \cdot \quad \cdot \quad (172)$$

$$\frac{dc''}{c''} = \frac{L_1 + c'L_2}{c' - c''} \frac{dT}{T^2\psi''} \quad \cdot \quad \cdot \quad \cdot \quad (173)$$

From this it follows immediately, that the concentrations c' and c'' also change in the same sense. For not only ψ' and ψ'' (§ 217), but also L_1 and L_2, the heat of vaporization of both components, are essentially positive.

We shall now suppose that $c'' > c'$. This in no way limits the generality, since we can always denote by the suffix 1 that constituent (*e.g.* water), which is present in a higher percentage in the single dashed phase (liquid) than in the double dashed phase (vapour). The suffix 2 then denotes that constituent (*e.g.* alcohol), which is present in a greater percentage in the double dashed phase than in the single dashed phase. This constituent is called the *more volatile* constituent. The last equations then tell us that with rise of temperature the concentration of the more volatile component diminishes simultaneously in both phases. Also, *e.g.*, by the continuous distillation of a mixture of water and alcohol, the distillate as well as the residue, with rising temperature, becomes poorer in alcohol. This arises from the fact that the ratio of the quantity of alcohol to that of water evaporating at any moment is greater than c', but smaller than c''. The first condition makes the liquid poorer in alcohol ($dc' < 0$), the second makes the vapour poorer in alcohol ($dc'' < 0$).

On continuing this isobaric evaporation, two cases may

arise, according as the concentration c', which lies between 0 and c'', finally coincides with 0 or with c''. In the first case ($c' = 0$) the more volatile component has finally passed completely into the second phase, and the first phase contains the less volatile constituent to any degree of purity. This occurs with water and ethyl alcohol. In the second case, $c' = c''$ at a definite temperature. As is to be seen from the last two equations, the boiling point does not alter with the concentration. Further evaporation takes place without rise of temperature and the mixture boils at constant temperature. The distillate and the residue continue to have the same percentage composition. A mixture of water and about 80% formic acid boils at constant temperature. This percentage depends somewhat on the pressure.

The boiling point T of a mixture with this particular proportion is a maximum : $\dfrac{dT}{dc'} = 0$. Each solution of formic acid in water, on distillation, approaches this mixture with rising temperature, whether we start with an acid content greater or smaller than 80%. In the first case, the acid content diminishes in both phases on continuous evaporation, in the second case it increases. According to (172) and (173), in the first case the acid is the more volatile constituent ($c'' > c'$) *i.e.* the vapour is richer in acid than the liquid. In the second case (acid content less than 80%), water is the more volatile constituent ($c' > c''$), *i.e.* the liquid is richer in acid than the vapour. With rising temperature both phases become richer in acid until finally the 80% content is reached.

The equation $\dfrac{dT}{dc'} = 0$ is also satisfied, if the boiling point relative to the concentration is a minimum, as, for example, with a mixture of water and propyl alcohol. According to equation (172), $c' = c''$, *i.e.* the mixture boils at constant temperature. This mixture is, however, unstable, since on changing the concentration by the smallest amount to one side or the other, continuous distillation increases the difference of the concentrations of both phases. Since then a rise of the boiling point accompanies the distillation, the boiling point moves away from its minimum value, and accordingly the concentrations from the composition of the constant boiling mixture.

§ **219A.** If, on the other hand, the evaporation of the liquid mixture takes place at constant temperature ($d\mathrm{T} = 0$) by a gradual increase of the volume, then with decreasing pressure ($dp < 0$) the concentrations, according to (170) and (171), change in the following way :

$$\frac{dc'}{c'} = \frac{v_1 + c''v_2}{c'' - c'} \cdot \frac{dp}{T\psi''}$$

$$\frac{dc''}{c''} = \frac{v_1 + c'v_2}{c'' - c'} \cdot \frac{dp}{T\psi''}.$$

Here also both concentrations c' and c'' change simultaneously in the same sense, since, besides ψ' and ψ'', also v_1 and v_2, the changes of volume of both constituents on evaporation are essentially positive. If, further, $c'' > c'$, according to the stipulation of the last paragraph, then on continuous evaporation with the consequent decrease of pressure both liquid and vapour become poorer in the more volatile constituent, until finally that constituent completely vanishes from the liquid phase ($c' = 0$) or the mixture reaches the composition ($c' = c''$). Further evaporation produces no change of pressure and no further change of concentration. These proportions correspond to those of evaporation under constant pressure.

§ **219B.** The equations (172) and (173), of course, hold also if the second phase is liquid or solid, *e.g.* for the freezing point of an alloy (*e.g.* bismuth and lead). It follows that the temperature of solidification of a liquid alloy rises, if the liquid alloy (the one dashed phase) is enriched by that constituent (2), which predominately precipitates. Since by supposition $c'' > c'$, and since L_1 and L_2 are both negative, then $\frac{d\mathrm{T}}{dc'} > 0$. According to the same principle, the freezing point of an aqueous salt solution rises when the solution is diluted with water, if pure ice freezes out. In the limit when the temperature of solidification does not alter with the concentration $\left(\dfrac{dT}{dc'} = 0\right)$, $c'' = c'$. The alloy solidifies without change of concentration. This alloy is called the *eutectic* alloy.

§ **220.** In the following applications we shall restrict ourselves to the case in which the second constituent occurs only in the first phase,

$$c'' = 0,$$

and, therefore, also $\quad dc'' = 0. \quad . \quad . \quad . \quad . \quad (174)$

The first constituent which occurs along with the second in the first phase, and pure in the second, will be called the *solvent;* the second, the *dissolved substance.* The equation (171) falls away, and from (170) there remains

$$\frac{L}{T^2}dT - \frac{v}{T}dp - \psi dc = 0, \quad . \quad . \quad . \quad (175)$$

if we omit suffixes and dashes for simplicity.

We shall take, first, a solution of a nonvolatile salt in contact with the vapour of the solvent, and investigate the equation (175) in three directions by keeping in turn the concentration c, the temperature T, and the pressure p constant.

§ **221. Concentration Constant:** $dc = 0.$—The relation between the vapour pressure and the temperature is, by (175),

$$\left(\frac{\partial p}{\partial T}\right)_c = \frac{L}{T . v} \quad . \quad . \quad . \quad . \quad (176)$$

Here L may be called briefly the heat of vaporization of the solution. If, instead of regarding L as the ratio of two infinitely small quantities, we take it to be the heat of vaporization per unit mass of the solvent, then the mass of the solvent must be assumed so large that the concentration is not appreciably altered by the evaporation of unit mass. v may generally be put equal to the specific volume of the vapour.

Assuming, further, that the laws of Boyle and Gay Lussac hold for the vapour, we get

$$v = \frac{R}{m} \cdot \frac{T}{p} \quad \ldots \ldots \quad (177)$$

and, by (176), $\qquad L = \frac{R}{m} T^2 \left(\frac{\partial \log p}{\partial T} \right)_c .$

On the other hand, L is also the quantity of heat given out when unit mass of the vapour of the solvent combines at constant temperature and pressure with a large quantity of a solution of concentration c. This process may be performed directly, or unit mass of the vapour may be first condensed to the pure solvent and then the solution diluted with it. If the initial and final states of the system are the same in both cases, then by the first law the sums of the heat evolved and the work spent are equal. The *heat of dilution* of a solution may be derived in this way.

In the first case the sum of the heat given out and the work spent is

$$L - pv = \frac{R}{m} T^2 \left(\frac{\partial \log p}{\partial T} \right)_c - pv.$$

In the second case, by the method used in § 215 we obtain, as the sum of the heat given out and the work spent during condensation and dilution,

$$\frac{R}{m} T^2 \frac{d \log p_0}{dT} - p_0 v_0 + \Delta,$$

where p_0 is the pressure, v_0 the specific volume of the vapour of the solvent in contact with the pure liquid solvent, Δ the heat of dilution of the solution, *i.e.* the heat given out on adding unit mass of the solvent to a large quantity of the solution of concentration c. Both the above expressions being equal according to the first law, we obtain, on applying Boyle's law,

$$\Delta = \frac{R}{m} T^2 \left(\frac{\partial \log \dfrac{p}{p_0}}{\partial T} \right)_c , \quad \ldots \quad (178)$$

which is Kirchhoff's formula for the heat of dilution.

The quantities here neglected, by considering the vapour a perfect gas, and its specific volume large in comparison with that of the liquid, may readily be taken into account when necessary.

The similarity of the expressions for Δ, the heat of dilution, and for λ, the heat of saturation (161), is only external, since in this case the solution may be of any concentration, and therefore may be differentiated with respect to the temperature, c being kept constant, while in (161) the concentration of a saturated solution changes with temperature in a definite manner.

§ **222.** Since Δ is small for *small* values of c (dilute solutions, § 97), then, according to (178), the ratio of the vapour pressure of a dilute solution of fixed concentration to the vapour pressure of the pure solvent is practically independent of the temperature (Babo's law).

§ **223. Temperature Constant:** $d\mathrm{T} = 0$.—The relation between the vapour pressure (p) and the concentration (c) of the solution is, according to (175),

$$\left(\frac{\partial p}{\partial c}\right)_{\mathrm{T}} = -\frac{\mathrm{T}\psi}{v} \quad . \quad . \quad . \quad . \quad (179)$$

Neglecting the specific volume of the liquid in comparison with that of the vapour, and considering the latter a perfect gas of molecular weight m, equation (177) gives

$$\left(\frac{\partial p}{\partial c}\right)_{\mathrm{T}} = -\frac{mp\psi}{\mathrm{R}},$$

or

$$\left(\frac{\partial \log p}{\partial c}\right)_{\mathrm{T}} = -\frac{m}{\mathrm{R}}\psi.$$

Since ψ is always positive (§ 217), the vapour pressure must decrease with increasing concentration. This proposition furnishes a means of distinguishing between a solution and an emulsion. In an emulsion the number of particles suspended in the solution has no influence on the vapour pressure.

So long as the quantity ψ is undetermined, nothing further can be stated with regard to the general relation between the vapour pressure and the concentration.

§ 224. As we have $p = p_0$ when $c = 0$ (pure solvent), $p - p_0$ is small for small values of c. We may, therefore, put

$$\frac{\partial p}{\partial c} = \frac{p - p_0}{c - 0} = \frac{p - p_0}{c}.$$

Hence, by (179),

$$p_0 - p = \frac{cT\psi}{v} \quad . \quad . \quad . \quad . \quad . \quad (180)$$

and substituting for v, as in (177), the specific volume of the vapour, considered a perfect gas, we get

$$\frac{p_0 - p}{p} = \frac{cm\psi}{R}. \quad . \quad . \quad . \quad (181)$$

This means that the relative decrease of the vapour pressure is proportional to the concentration of the solution (Wüllner's law). For further particulars, see § 270.

§ 225. Pressure Constant: $dp = 0$.—The relation between the temperature (boiling point) and the concentration is, by (175),

$$\left(\frac{\partial T}{\partial c}\right)_p = \frac{T^2\psi}{L} \quad . \quad . \quad . \quad . \quad (182)$$

Since ψ is positive, the boiling point rises with increasing concentration. By comparing this with the formula (179) for the decrease of the vapour pressure, we find that any solution gives

$$\left(\frac{\partial T}{\partial c}\right)_p : \left(\frac{\partial p}{\partial c}\right)_T = -\frac{Tv}{L},$$

i.e. for an infinitely small increase of the concentration the rise in the boiling point (at constant pressure) is to the decrease of the vapour pressure (at constant temperature) as the product of the absolute temperature and the specific volume of the vapour is to the heat of vaporization of the solution.

Remembering that this relation satisfies the identity

$$\left(\frac{\partial T}{\partial c}\right)_p : \left(\frac{\partial p}{\partial c}\right)_T = -\left(\frac{\partial T}{\partial p}\right)_c,$$

we come immediately to the equation (176).

§ 226. Let T_0 be the boiling point of the pure solvent ($c = 0$), then, for small values of c, the difference between T and T_0 will be small, and we may put

$$\frac{\partial T}{\partial c} = \frac{T - T_0}{c - 0} = \frac{T - T_0}{c},$$

whereby the equation becomes

$$T - T_0 = \frac{cT^2\psi}{L}. \quad . \quad . \quad . \quad . \quad . \quad . \quad (183)$$

This means that the elevation of the boiling point is proportional to the concentration of the solution. For further details, see § 269.

§ 227. Let the second phase consist of the pure solvent in the solid state instead of the gaseous state, as happens in the freezing of an aqueous salt solution or in the precipitation of salt from a saturated solution. In the latter case, in conformity with the stipulations of § 220, the salt will be regarded as the first constituent (the solvent), and water as the second constituent (the dissolved substance). The equation (175) is then directly applicable, and may be discussed in three different ways. We may ask how the freezing point or the saturation point of a solution of definite concentration changes with the pressure ($dc = 0$); or, how the pressure must be changed, in order that a solution of changing concentration may freeze or become saturated at constant temperature ($dT = 0$); or, finally, how the freezing point or the saturation point of a solution under given pressure changes with the concentration ($dp = 0$). In the last and most important case, if we denote the freezing point or the saturation point as a function of the concentration by T′, to distinguish it from the boiling point, equation (175) gives

$$\left(\frac{\partial T'}{\partial c}\right)_p = \frac{T^2\psi}{L'}. \quad \ldots \quad (184)$$

L′ being the heat absorbed when unit mass of the solvent separates as a solid (ice, salt) from a large quantity of the solution of concentration c. This heat quantity is usually negative. It is called the *heat of solidification* of the solution or the *heat of precipitation* of the salt.

The heat of solidification (L′) of a salt solution is always negative, hence the freezing point is lowered by an increase of concentration c. On the other hand, if the heat of precipitation (L′) of a salt from a solution be negative, the saturation point T′ is lowered by an increase of the water content, c, of the solution, or rises with an increase of the concentration of the salt. If L′ be positive, the saturation point is lowered by an increase of the concentration of the salt. Should we prefer to designate by c, not the amount of water, but the amount of salt in a saturated solution, then, according to the definition of c in (162) and of ψ in (165), we should have $\frac{1}{c}$ replacing c in (184) and $c\psi$ replacing ψ, and therefore

$$\left(\frac{\partial T'}{\partial c}\right)_p = -\frac{T^2\psi}{cL'}. \quad \ldots \quad (185)$$

Here c and ψ have the same meaning as in equation (184), which refers to the freezing point of a solution.

§ 228. Let T_0' be the freezing point of the pure solvent ($c = 0$), then, for small values of c, T′ will be nearly $= T_0'$ and we may put

$$\frac{\partial T'}{\partial c} = \frac{T' - T_0'}{c - 0} = \frac{T' - T_0'}{c}.$$

Equation (184) then becomes

$$T' - T_0' = \frac{cT^2\psi}{L'} \quad \ldots \quad \ldots \quad (186)$$

which means that the lowering of the freezing point is proportional to the concentration. (For further particulars see § 269.)

§ 229. The positive quantity, ψ, which occurs in all these formulæ, has a definite value for a solution of given c, T, and p, and is independent of the nature of the second phase. Our last equations, therefore, connect in a perfectly general way the laws regarding the lowering of the vapour pressure, the elevation of the boiling temperature, the depression of the freezing point, and the change of the saturation point. Only one of these phenomena need be experimentally investigated in order to calculate ψ, and by means of the value thus determined the others may be deduced for the same solution.

We shall now consider a further case for which ψ is of fundamental importance, viz. the state of equilibrium which ensues when the pure liquid solvent forms the second phase, not in contact with a solution, for no equilibrium would thus be possible, but separated from it by a membrane, permeable to the solvent only. It is true that for no solution can perfectly *semipermeable* membranes of this character be manufactured. In fact, the further development of this theory (§ 259) will exclude them as a matter of principle, for in every case the dissolved substance will also diffuse through the membrane, though possibly at an extremely slow rate. For the present it is sufficient that we may, without violating a law of thermodynamics, assume the velocity of diffusion of the dissolved substance as small as we please in comparison with that of the solvent. This assumption is justified by the fact that semipermeability may be very closely approximated in the case of many substances. The error committed in putting the rate of diffusion of a salt through such a membrane equal to zero, falls below all measurable limits. An exactly similar error is made in assuming that a salt does not evaporate or freeze from a solution, for, strictly speaking, this assumption is not admissible (§ 259).

The condition of equilibrium of two phases separated by a

semipermeable membrane is contained in the general thermodynamical condition of equilibrium (145),

$$\delta\Psi' + \delta\Psi'' = 0, \quad . \quad . \quad . \quad . \quad (187)$$

which holds for virtual changes at constant temperature
and pressure in each phase. The only difference between
this case and free contact is, that the pressures in the two
phases may be different. Pressure always means hydrostatic
pressure as measured by a manometer. If, in the general
equation (76), we put

$$W = -p'\delta V' - p''\delta V'',$$

it immediately follows that (187) is the condition of equilibrium. The further conclusions from (187) are completely
analogous to those which are derived, when there is a free
surface of contact. Corresponding to (163) we have for any
displacement of the equilibrium

$$\frac{Q}{T^2}dT - \frac{\delta V'}{T}dp' - \frac{\delta V''}{T}dp'' + dM_1'\delta\frac{\partial\Psi'}{\partial M_1'} + dM_2'\delta\frac{\partial\Psi'}{\partial M_2'} + \ldots = 0.$$

Since the constituent 2 occurs only in the first phase, we get,
instead of (175),

$$\frac{L}{T^2}dT - \frac{v'}{T}dp' - \frac{v''}{T}dp'' - \psi dc = 0. \quad . \quad . \quad (188)$$

Here, as in § 221, L is the " *heat of removal* " of the solvent
from the solution, *i.e.* the heat absorbed when, at constant
temperature and constant pressures p' and p'', unit mass of
the solvent passes through the semipermeable membrane
from a large quantity of the solution to the pure solvent.
The change of volume of the solution during this process
is v' (negative), that of the pure solvent v'' (positive). In the
condition of equilibrium (188), three of the four variables
T, p', p'', c remain arbitrary, and the fourth is determined by
their values.

Consider the pressure p'' in the pure solvent as given and
constant, say one atmosphere, then $dp'' = 0$. If, further, we
put $dT = 0$ and dc not equal to zero, we are then considering

solutions in which the concentration varies, but the temperature and the pressure in the pure solvent remains the same. Then, by (188),

$$\left(\frac{\partial p'}{\partial c}\right)_{\mathrm{T}} = -\frac{\mathrm{T}\psi}{v'}.$$

Since $\psi > 0$, and $v' < 0$, p' the pressure in the solution increases with the concentration.

The difference of the pressures in the two phases, $p' - p'' = \mathrm{P}$, has been called the *osmotic pressure* of the solution. Since p'' has been assumed constant, we may write

$$\left(\frac{\partial \mathrm{P}}{\partial c}\right)_{\mathrm{T}} = -\frac{\mathrm{T}\psi}{v'} \quad \ldots \quad (189)$$

Accordingly, the osmotic pressure increases with increasing concentration. Since $p' - p$ vanishes when $c = 0$, the osmotic pressure is always positive.

For small values of c,

$$\frac{\partial \mathrm{P}}{\partial c} = \frac{\mathrm{P} - 0}{c - 0} = \frac{\mathrm{P}}{c}$$

and $- v'$ is nearly equal to the specific volume of the solution. It therefore follows from (189) that

$$\mathrm{P} = \frac{c\mathrm{T}\psi}{v}, \quad \ldots \quad \ldots \quad (190)$$

where v denotes the specific volume of the solution. A further discussion of this question will be found in § 272.

Thus the laws of the osmotic pressure have also been expressed in terms of ψ, which controls those of the depression of the freezing point, the elevation of the boiling point, etc. Or, we may say, that the last named laws can all be expressed in terms of the osmotic pressure, for ψ can be eliminated from equation (189) and one of the above equations. It is particularly noteworthy that the relation which has been deduced is independent of all molecular assumptions and presentations, although these have played an important *rôle* in the development of the theory.

§ **230.** We have expressed the laws of equilibrium of several systems that fulfil the conditions of § 220, in terms of a quantity ψ, which is characteristic for the thermodynamical behaviour of a solution. There is no difficulty in deducing the corresponding laws when the dissolved substance is contained also in the second phase. Starting from the two equations (170) and (171), we find that all the relations in question depend on ψ' and ψ''. A better insight into the nature of these quantities is gained by extending to the liquid state the idea of the molecule, hitherto applied only to the gaseous state. This step is taken in the next two chapters, and it appears that the manner in which this idea applies is uniquely determined by the laws of thermodynamics which have been given.

§ **231.** Just as the conditions of equilibrium (170) and (171) for two independent constituents in two phases were deduced from the general relation (153), so in the same way a similar deduction may be made in the general case.

We shall conclude this chapter by giving, briefly, the results for a system of α independent constituents in β phases.

Denoting the concentrations of the independent constituents, relative to one fixed constituent 1, by

$$\frac{M_2'}{M_1'} = c_2'; \quad \frac{M_3'}{M_1'} = c_3'; \quad \frac{M_4'}{M_1'} = c_4'; \quad \ldots$$

$$\frac{M_2''}{M_1''} = c_2''; \quad \frac{M_3''}{M_1''} = c_3''; \quad \frac{M_4''}{M_1''} = c_4''; \quad \ldots$$

$$\cdots \cdots \cdots \cdots \cdots \cdots \cdots \cdots \cdots$$

the condition that, by any infinitely small change of the system : $d\mathrm{T}$, dp, dc_2', dc_3', dc_4', $\ldots$ dc_2'', dc_3'', dc_4'', $\ldots$, the equilibrium may remain stable with regard to the passage of the constituent 1 from the phase denoted by one dash to the phase denoted by two dashes is

$$\frac{L_1}{T^2}d\mathrm{T} - \frac{v_1}{T}dp + (\psi_2''dc_2'' - \psi_2'dc_2') + (\psi_3''dc_3'' - \psi_3'dc_3') + \ldots = 0,$$

where, analogous to (165),

$$\psi_2' = M_1' \frac{\partial^2 \Psi'}{\partial M_1' \partial M_2'}; \quad \psi_3' = M_1' \frac{\partial^2 \Psi'}{\partial M_1' \partial M_3'}; \ \cdots$$

$$\psi_2'' = M_1'' \frac{\partial^2 \Psi''}{\partial M_1'' \partial M_2''}; \quad \psi_3'' = M_1'' \frac{\partial^2 \Psi''}{\partial M_1'' \partial M_3''}; \ \cdots$$

and L_1, v_1 denote the heat absorbed, and the increase of volume of the system during the isothermal and isobaric transference of unit mass of constituent 1 from a large quantity of the phase denoted by one dash to a large quantity of the phase denoted by two dashes (cf. § 221).

The corresponding conditions of equilibrium for any possible passage of any constituent from any one phase to any other phase may be established in the same way.

CHAPTER IV.

GASEOUS SYSTEM.

§ **232.** The relations, which have been deduced from the general condition of equilibrium (79) for the different properties of thermodynamical equilibria, rest mainly on the relations between the characteristic function Ψ, the temperature, and the pressure as given in the equations (79b). It will be impossible to completely answer all questions regarding equilibrium until Ψ can be expressed in its functional relation to the masses of the constituents in the different phases. The introduction of the molecular weight serves this purpose. Having already defined the molecular weight of a chemically homogeneous gas as well as the number of molecules of a mixture of gases by Avogadro's law, we shall turn first to the investigation of a system consisting of one gaseous phase.

The complete solution of the problem consists in expressing Ψ in terms of T, p, and n_1, n_2, n_3, ..., the number of all the different kinds of molecules in the mixture.

Since we have, in general, by (75),

$$\Psi = \Phi - \frac{U + pV}{T},$$

we are required to express the entropy Φ, the energy U, and the volume V as functions of the above independent variables. This can be done, in general, on the assumption that the mixture obeys the laws of perfect gases. Such a restriction will not, in most cases, lead to appreciable errors. Even this assumption may be set aside by special measurement of the quantities Φ, U, and V. For the present, however, perfect gases will be assumed.

§ **233.** The laws of Boyle, Gay Lussac, and Dalton determine the volume of the mixture, for equation (16) gives

$$V = \frac{RT}{p}(n_1 + n_2 + \ldots) = \frac{RT}{p}\sum n_1 . \quad . \quad (191)$$

By the first law of thermodynamics, the energy U of a mixture of gases is given by the energies of its constituents, for, according to this law, the energy of the system remains unchanged, no matter what internal changes take place, provided there are no external effects. Experience shows that when diffusion takes place between a number of gases at constant temperature and pressure, neither does the volume change, nor is heat absorbed or evolved. The energy of the system, therefore, remains constant during the process. Hence, the energy of a mixture of perfect gases is the sum of the energies of the gases at the same temperature and pressure. Now the energy U_1 of n_1 molecules of a perfect gas depends only on the temperature; it is, by (35),

$$U_1 = n_1(C_{v_1}T + b_1), \quad . \quad . \quad . \quad (192)$$

where C_{v_1} is the molecular heat of the gas at constant volume, and b_1 is a constant. Hence the total energy of the mixture is

$$U = \sum n_1(C_v T + b_1) \quad . \quad . \quad . \quad (193)$$

§ **234.** We have now to determine the entropy Φ as a function of T, p, and n_1, n_2, ... the number of molecules. Φ, in so far as it depends on T and p, may be calculated from the equation (60),

$$d\Phi = \frac{dU + pdV}{T},$$

where the differentials correspond to variations of T and p, but not of the number of molecules.

Now, by (193),

$$dU = \sum n_1 C_{v_1} dT$$

and, by (191),

$$dV = R\sum n_1 d\left(\frac{T}{p}\right)$$

therefore, $\quad d\Phi = \sum n_1\left(C_{v_1}\frac{dT}{T} + \frac{RdT}{T} - \frac{Rdp}{p}\right),$

or, since $C_{v_1} + R = C_{p_1}$,

$$d\Phi = \sum n_1\left(C_{p_1}\frac{dT}{T} - R\frac{dp}{p}\right)$$

and, by integration,

$$\Phi = \sum n_1(C_{p_1}\log T - R\log p + k_1) + C. \quad (194)$$

Besides the constants $k_1, k_2, k_3 \ldots$ which depend on the nature of the individual gases, and on the units of T and p, another constant C is added. The constants $k_1, k_2, \ldots$ give a linear relation between the entropy Φ and the number of the molecules $n_1, n_2 \ldots$, while the constant of integration may, and will be found to, depend in a more complicated way on the composition of the mixture, *i.e.* on the ratios of the number of molecules. The investigation of this relation forms the most important part of our problem.

The determination of the constant is not, in this case, a matter of definition. It can only be determined by applying the second law of thermodynamics to a reversible process which brings about a change in the composition of the mixture. A reversible process produces a definite change of the entropy which may be compared with the simultaneous changes of the number of molecules, and thus the relation between the entropy and the composition determined. If we select a process devoid of external effects either in work or heat, then the entropy remains constant during the whole process. We cannot, however, use the process of diffusion, which leads to the value of U; for diffusion, as might be expected, and as will be shown in § 238, is irreversible, and therefore leads only to the conclusion that the entropy of the system is thereby increased. There is, however, a reversible process at our disposal, which will change the composition of the mixture, viz. the separation by a semipermeable membrane, as introduced and established in § 229.

§ 235. Before we can apply a semipermeable membrane to the purpose in hand, we must acquaint ourselves with the nature of the thermodynamical equilibrium of a gas in contact with both sides of a membrane permeable to it. The membrane will act like a bounding wall to those gases to which it is impermeable, and will, therefore, not introduce any special conditions.

Experience shows that a gas on both sides of a membrane permeable to it is in equilibrium when its partial pressures (§ 18) are the same on both sides, quite independent of the other gases present. This proposition is neither axiomatic nor a necessary consequence of the preceding considerations, but it commends itself by its simplicity, and has been confirmed without exception in the few cases accessible to direct experiment.

A test of this kind may be established as follows : Platinum foil at a white heat is permeable to hydrogen, but impermeable to air. If a vessel having a platinum wall be filled with pure hydrogen, and hermetically sealed, the hydrogen must completely diffuse out against atmospheric pressure when the platinum is heated. As the air cannot enter, the vessel must finally become completely exhausted.*

§ 236. We shall make use of the properties of semi-per-

* This inference was tested by me in the Physical Institute of the University of Munich in 1883, and was confirmed within the limits of experimental error as far as the unavoidable deviations from ideal conditions might lead one to expect. As this experiment has not been published anywhere, I shall briefly describe it here. A glass tube of about 5 mm. internal diameter, blown out to a bulb at the middle, was provided with a stop-cock at one end. To the other end a platinum tube 10 cm. long was fastened, and closed at the end. The whole tube was exhausted by a mercury pump, filled with hydrogen at ordinary atmospheric pressure, and then closed. The closed end of the platinum portion was then heated in a horizontal position by a Bunsen burner. The sealing wax connection between the glass and platinum tubes had to be kept cool by a continuous current of water to prevent the softening of the wax. After four hours the tube was taken from the flame, cooled to the temperature of the room, and the stop-cock opened under mercury. The mercury rose rapidly almost completely filling the tube, proving that the tube was to a certain extent exhausted.

meable membranes to separate in a reversible and simple
manner the constituents of a gas mixture. Let us consider
the following example :—

Let there be four pistons in a hollow cylinder, two of
them, A and A′, in fixed positions; two, B and B′, movable
in such a way that the distance BB′ remains constant, and
equal to AA′. This is indicated by the brackets in Fig. 5.
Further, let A′ (the bottom), and B (the cover) be imper-

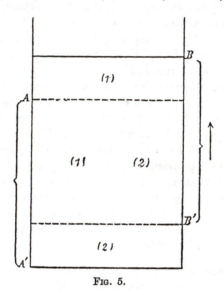

Fig. 5.

meable to any gas, while A is permeable only to one gas
(1), and B′ only to another one (2). The space above B is a
vacuum.

At the beginning of the process the piston B is close to
A, therefore B′ close to A′, and the space between them
contains a mixture of the two gases (1 and 2). The connected pistons B and B′ are now very slowly raised. The
gas 1 will pass into the space opening up between A and B,
and the gas 2 into that between A′ and B′. Complete separation will have been effected when B′ is in contact with A.

We shall now calculate the external work of this process. The pressure on the movable piston B consists only of the pressure of the gas 1, upwards, since there is a vacuum above B; and on the other movable piston, B', there is only the partial pressure of the same gas, which acts downwards. According to the preceding paragraph both these pressures are equal, and since the paths of B and B' are also equal, the total work done on the pistons is zero. If no heat be absorbed or given out, as we shall further assume, the energy of the system remains constant. But, by (193), the energy of a mixture of gases depends, like that of pure gases, on the temperature alone, so the temperature of the system remains constant throughout.

Since this infinitely slow process is reversible, the entropy in the initial and final states is the same, if there are no external effects. Hence, the entropy of the mixture is equal to the sum of the entropies which the two gases would have, if at the same temperature each by itself occupied the whole volume of the mixture. This proposition may be easily extended to a mixture of any number of gases. *The entropy of a mixture of gases is the sum of the entropies which the individual gases would have, if each at the same temperature occupied a volume equal to the total volume of the mixture.* This proposition was first established by Gibbs.

§ **237.** The entropy of a perfect gas of mass M and molecular weight m was found to be (52)

$$M\left(\frac{C_v}{m}\log T + \frac{R}{m}\log v + \text{const.}\right),$$

where C_v is the molecular heat at constant volume, as in (192). By the gas laws (14), the volume of unit mass is

$$v = \frac{R}{m} \cdot \frac{T}{p},$$

whence the entropy is

$$n(C_v \log T + R \log \frac{T}{p} + k),$$
$$= n(C_p \log T - R \log p + k), \quad . \quad . \quad . \quad (195)$$

where $n = \dfrac{M}{m}$, the number of molecules, and the constant k includes the term $\log \dfrac{R}{m}$. Hence, according to Gibbs's proposition, the entropy of the mixture is

$$\Phi = \sum n_1 (C_{p_1} \log T - R \log p_1 + k_1),$$

p_1 being the partial pressure of the first gas in the mixture.

Now, by (8), the pressure of the mixture is the sum of the partial pressures, $\Sigma p_1 = p$, and, by § 40, the partial pressures are proportional to the number of molecules of each gas.

$$p_1 : p_2 : \ldots = n_1 : n_2 : \ldots$$

Hence
$$p_1 = \frac{n_1}{n_1 + n_2 + \ldots} p$$

$$p_2 = \frac{n_2}{n_1 + n_2 + \ldots} p$$

$$\cdot \quad \cdot \quad \cdot \quad \cdot \quad \cdot \quad \cdot$$

or, if we introduce the *concentrations* of the different gases in the mixture,

$$c_1 = \frac{n_1}{n_1 + n_2 + \ldots}; \quad c_2 = \frac{n_2}{n_1 + n_2 + \ldots};$$

$$p_1 = c_1 p; \qquad\qquad p_2 = c_2 p \quad . \quad . \quad . \quad (196)$$

Thus the expression for the entropy of a mixture as a function of T, p, and n the number of molecules, finally becomes

$$\Phi = \sum n_1 (C_{p_1} \log T - R \log (c_1 p) + k_1) \quad . \quad (197)$$

Comparing this expression with the value of the entropy of a gas mixture given by (194), we see that the constant of integration which was left undetermined is

$$C = - R \sum n_1 \log c_1 \quad . \quad . \quad . \quad (198)$$

§ 238. Knowing the value of the entropy of a gas mixture, we may answer the question which we discussed in § 234, whether and to what extent the entropy of a system of gases is increased by diffusion. Let us take the simplest case, that of two gases, the number of molecules being n_1 and n_2, diffusing into one another under common and constant pressure and temperature. Before diffusion begins, the entropy of the system is the sum of the entropies of the gases, by (195),

$$n_1(C_{p_1} \log T - R \log p + k_1) + n_2(C_{p_2} \log T - R \log p + k_2).$$

After diffusion it is, by (197),

$$n_1(C_{p_1} \log T - R \log (c_1 p) + k_1)$$
$$+ n_2(C_{p_2} \log T - R \log (c_2 p) + k_2).$$

Therefore, the change of the entropy of the system is

$$- n_1 R \log c_1 - n_2 R \log c_2,$$

by (196), an essentially positive quantity. This shows that diffusion is always irreversible.

It also appears that the increase of the entropy depends solely on the number of the molecules n_1, n_2, and not on the nature, *e.g.* the molecular weight, of the diffusing gases. The increase of the entropy does not depend on whether the gases are chemically alike or not. By making the two gases the same, there is evidently no increase of the entropy, since no change of state ensues. It follows that the chemical difference of two gases, or, in general, of two substances, cannot be represented by a continuous variable; but that here we can speak only of a discontinuous relation, either of equality or inequality. This fact involves a fundamental distinction between chemical and physical properties, since the latter may always be regarded as continuous (cf. Note to § 35).

§ 239. The values of the entropy (197), the energy (193), and the volume (191), substituted in (75) give the function Ψ.

$$\Psi = \sum n_1(C_{p_1} \log T - R \log (c_1 p) + k_1 - C_{v_1} - \frac{b_1}{T} - R),$$

or, if, for shortness, we put

$$k_1 - C_{v_1} - R = k_1 - C_{p_1} = a_1, \quad . \quad . \quad (198a)$$

and the quantity, which depends on T and p and not on the number of molecules

$$C_{p_1} \log T - \frac{b_1}{T} - R \log p + a_1 = \psi \ . \qquad (199)$$

$$\Psi = \sum n_1(\psi_1 - R \log c_1) \ . \quad . \quad . \ (199a)$$

§ 240. This enables us to establish the condition of equilibrium. If in a gas mixture a chemical change, which changes the number of molecules $n_1 \, n_2 \ldots$ by $\delta n_1, \ \delta n_2 \ldots$ be possible, then such a change will not take place if the condition of equilibrium (79) be fulfilled, *i.e.* if, when $\delta T = 0$ and $\delta p = 0$,

$$\delta \Psi = 0,$$

or $\quad \sum(\psi_1 - R \log c_1)\delta n_1 + \sum n_1 \delta(\psi_1 - R \log c_1) = 0 \ \ (200)$

The quantities $\psi_1, \ \psi_2 \ldots$ depend on T and p only, therefore

$$\delta\psi_1 = \delta\psi_2 = \ldots = 0.$$

Further,

$$n_1 \delta \log c_1 + n_2 \delta \log c_2 + \ldots = \frac{n_1}{c_1}\delta c_1 + \frac{n_2}{c_2}\delta c_2 + \ldots$$

and, by (196),

$$= (n_1 + n_2 + \ldots)(\delta c_1 + \delta c_2 + \ldots) = 0,$$

since $\qquad c_1 + c_2 + \ldots = 1.$

The condition of equilibrium, therefore, reduces to

$$\sum(\psi_1 - R \log c_1)\delta n_1 = 0.$$

Since this equation does not involve the absolute values of the variations δn_1, but only their ratios, we may put

$$\delta n_1 : \delta n_2 : \ldots = \nu_1 : \nu_2 : \ldots \quad . \quad . \ (201)$$

and take $\nu_1, \ \nu_2 \ldots$ to denote the number of molecules simultaneously passing into the mass of each constituent. They are simple integers, positive or negative, according as the gas in question is forming, or is being used up in the formation of others. The condition of equilibrium now becomes

$$\sum(\psi_1 - R \log c_1)\nu_1 = 0,$$

or $\nu_1 \log c_1 + \nu_2 \log c_2 + \ldots = \dfrac{\nu_1\psi_1 + \nu_2\psi_2 + \cdots}{R} = \log K.$

The right-hand side of the equation depends only on temperature and pressure (199). The equation gives a definite relation between the concentrations of the different kinds of molecules for given temperature and pressure.

§ 241. We shall now substitute the values of ψ_1, ψ_2, $\ldots$ If, for shortness, we put the constants

$$\nu_1 + \nu_2 + \nu_3 + \ldots = \nu, \quad \ldots \quad (201a)$$

$$\frac{\nu_1 a_1 + \nu_2 a_2 + \cdots}{R} = \log A, \quad \ldots \quad (201b)$$

$$\frac{\nu_1 b_1 + \nu_2 b_2 + \cdots}{R} = B \quad \ldots \quad (202)$$

$$\frac{\nu_1 C_{p_1} + \nu_2 C_{p_2} + \cdots}{R} = C \quad \ldots \quad (203)$$

then from (199) the condition of equilibrium becomes

$\nu_1 \log c_1 + \nu_2 \log c_2 + \ldots$

$$= \log A + C \log T - \frac{B}{T} - \nu \log p = \log K$$

or

$$c_1{}^{\nu_1} c_2{}^{\nu_2} \ldots = A e^{-\frac{B}{T}} T^C p^{-\nu} = K \quad \ldots \quad (203a)$$

§ 242. This condition may be further simplified by making use of the experimental fact (§ 50) that the atomic heat of an element remains unchanged in its combinations. By equation (203), the product RC denotes the change, which the sum of the molecular heats of all the molecules, or the heat capacity of the whole system under constant pressure, $n_1 C_{p_1} + n_2 C_{p_2} + \ldots$, undergoes during the chemical reaction. If the molecular heat of each molecule at constant volume is equal to the sum of the atomic heats at constant volume, then, according to the above proposition, the heat capacity of the whole system at constant volume would be, as the sum of all the atomic heats, unchanged. Then

$$\nu_1 C_{v_1} + \nu_2 C_{v_2} + \ldots = 0,$$

and, therefore, $C = \nu_1 + \nu_2 + \ldots = \nu.$

It appears, however, that this supposition is not fulfilled with sufficient approximation to justify its introduction.

§ **243.** According to equation (203*a*) the influence of the pressure on the equilibrium depends entirely on the number $\Sigma\nu_1$, which gives the degree to which the total number of molecules, therefore also the volume of the mixture, is increased by the reaction considered. Where the volume remains unchanged, as, *e.g.*, in the dissociation of hydriodic acid, considered below, the equilibrium is independent of the pressure.

The influence of the temperature depends further on the constants B and C, which are closely connected with the heat of the reaction. For, by the first law,

$$Q = \delta U + p\delta V,$$

which, by (193) and (191), T and p being constant, becomes

$$Q = \sum(C_{v_1}T + b_1 + RT)\delta n_1 = \sum(C_{p_1}T + b_1)\delta n_1.$$

If we express the heat of the reaction in terms of the finite numbers ν, instead of the infinitely small numbers δn, then the heat absorbed is :

$$L = \sum(C_{p_1}T + b_1)\nu_1,$$

and by (202) and (203),

$$L = R(B + CT)$$

or $\qquad L = 1\cdot985(B + CT)$ cal.

The heat of a chemical transformation in a system, which is before and after the transformation completely gaseous, is independent of the pressure and is a linear function of the temperature.

§ **244.** Before proceeding to numerical applications, we shall enumerate the principal equations.

Suppose that in a gaseous system

$$n_1 \, m_1; \quad n_2 \, m_2; \quad n_3 \, m_3; \ldots$$

(n the number of molecules, m the molecular weight) any chemical change be possible, in which the simultaneous changes of the number of molecules are

$$\delta n_1 : \delta n_2 : \delta n_3 : \ldots = \nu_1 : \nu_2 : \nu_3 : \ldots$$

(ν simple, positive or negative, integers) then there will be equilibrium, if the concentrations

$$c_1 = \frac{n_1}{n_1 + n_2 +} \; ; \quad c_2 = \frac{n_2}{n_1 + n_2 +} \; ; \ldots$$

satisfy the condition

$$c_1{}^{\nu_1} c_2{}^{\nu_2} c_3{}^{\nu_3} \ldots = A e^{-\frac{B}{T}} T^C p^{-\nu} \quad . \quad . \quad . \quad (204)$$

The heat absorbed during the change at constant temperature and pressure is

$$L = 1 \cdot 985 (B + CT) \text{ cal.} \quad . \quad . \quad . \quad (205)$$

while the change of volume is

$$v = R \nu \frac{T}{p} \quad . \quad . \quad . \quad . \quad (206)$$

§ **245. Dissociation of Hydriodic Acid.**—Since hydriodic acid gas splits partly into hydrogen and iodine vapour, the system is represented by three kinds of molecules :

$$n_1 \text{ HI}; \quad n_2 \text{ H}_2; \quad n_3 \text{ I}_2;$$

The concentrations are :

$$c_1 = \frac{n_1}{n_1 + n_2 + n_3}; \quad c_2 = \frac{n_2}{n_1 + n_2 + n_3}; \quad c_3 = \frac{n_3}{n_1 + n_2 + n_3}.$$

The reaction consists in the transformation of two molecules of HI into one of H_2 and one of I_2 :

$$\nu_1 = -2; \quad \nu_2 = 1; \quad \nu_3 = 1; \quad \nu = \nu_1 + \nu_2 + \nu_3 = 0.$$

By (204), therefore, in the state of equilibrium,

$$c_1^{-2}c_2^{1}c_3^{1} = Ae^{-\frac{B}{T}} T^{C}$$

or
$$\frac{c_2\,c_3}{c_1^{2}} = \frac{n_2\,n_3}{n_1^{2}} = Ae^{-\frac{B}{T}} T^{C}. \quad . \quad (207)$$

Since the total number of atoms of hydrogen $(n_1 + 2n_2)$ and of iodine $(n_1 + 2n_3)$ in the system are supposed to be known, equation (207) is sufficient for the determination of the three quantities, n_1, n_2, and n_3, at any given temperature. The pressure has in this case no influence on the equilibrium. This has been confirmed by the recent measurement of M. Bodenstein. The measurements of the degree of dissociation at three different temperatures are sufficient for the determination of the constant A, B, and C.

Thus the equilibrium of any mixture of hydriodic acid, hydrogen, and iodine vapour at any temperature, even when the hydrogen and the iodine are not present in equivalent quantities, is determined by (207). Equation (205) gives the heat of dissociation of two molecules of hydriodic acid into a molecule of hydrogen and a molecule of iodine vapour at any temperature.

§ 246. Dissociation of Iodine Vapour.—At high temperatures iodine vapour appreciably decomposes, leading to a system of two kinds of molecules :

$$n_1\,I_2; \; n_2\,I.$$

The concentrations are :

$$c_1 = \frac{n_1}{n_1 + n_2}; \; c_2 = \frac{n_2}{n_1 + n_2}.$$

The reaction consists in a splitting of the molecule I_2 into two molecules I,

$$\therefore \; \nu_1 = -1; \; \nu_2 = 2; \; \nu = \nu_1 + \nu_2 = 1,$$

and in equilibrium, by (204),

$$c_1^{-1}c_2^{2} = \frac{n_2^{2}}{n_1(n_1 + n_2)} = A'e^{-\frac{B'}{T}} \frac{T^{C'}}{p} . \quad . \quad (208)$$

§ 247. Graded Dissociation.—Since, by equation (208), the concentration c_2 of the monatomic iodine molecules does not vanish even at low temperatures, the decomposition of the iodine vapour should be taken into account in determining the dissociation of hydriodic acid. This will have practically no influence on the results of § 245, but nevertheless we shall give the more rigorous solution on account of the theoretical interest which attaches to it.

There are now four kinds of molecules in the system :

$$n_1 \, HI; \; n_2 \, H_2; \; n_3 \, I_2; \; n_4 \, I.$$

The concentrations are :

$$c_1 = \frac{n_1}{n_1 + n_2 + n_3 + n_4}, \quad c_2 = \frac{n_2}{n_1 + n_2 + n_3 + n_4},$$

$$c_3 = \frac{n_3}{n_1 + n_2 + n_3 + n_4}, \quad c_4 = \frac{n_4}{n_1 + n_2 + n_3 + n_4}$$

Two kinds of chemical changes are possible :

$$(1) \; \nu_1 = -2; \; \nu_2 = 1; \; \nu_3 = 1; \; \nu_4 = 0; \; \nu = 0;$$
$$(2) \; \nu_1' = 0; \; \nu_2' = 0; \; \nu_3' = -1; \; \nu_4' = 2; \; \nu' = 1.$$

There will be equilibrium for each of these, if, by (204),

$$(1) \; c_1^{\nu_1} c_2^{\nu_2} c_3^{\nu_3} c_4^{\nu_4} = \frac{c_2 c_3}{c_1^2} = \frac{n_2 n_3}{n_2^2} = A e^{-\frac{B}{T}} T^C$$

and

$$(2) \; c_1^{\nu_1'} c_2^{\nu_2'} c_3^{\nu_3'} c_4^{\nu_4'} = \frac{c_4^2}{c_3} = \frac{n_4^2}{n_3(n_1 + n_2 + n_3 + n_4)} = A' e^{-\frac{B'}{T}} \frac{T^{C'}}{p}.$$

The total number of hydrogen atoms $(n_1 + 2n_2)$ and of iodine atoms $(n_1 + 2n_3 + n_4)$ being known, we have four equations for the unique determination of the four quantities n_1, n_2, n_3, n_4.

§ 248. The general equation of equilibrium (204) also shows that at finite temperatures and pressures none of the concentrations, c, can ever vanish; in other words, that the dissociation can never be complete, nor can it completely vanish. There is always present a finite, though perhaps a very small number of all possible kinds of molecules. Thus, in water vapour at any temperature at least a trace of oxygen and hydrogen must be present (see also § 259). In a great number of phenomena, however, these quantities are too small to be of any importance.

CHAPTER V.

DILUTE SOLUTIONS.

§ 249. To determine Ψ as a function of the temperature T, the pressure p, and the number n of the different kinds of molecules in a system of any number of constituents and any number of phases, we may use the method of the preceding chapter. It is necessary first to find by suitable measurements the volume V, and the internal energy U of each single phase, and then calculate the entropy Φ from the definition (60). A simple summation extending over all the phases gives, by (75), the function Ψ for the whole system. On account of incomplete experimental data, however, the calculation of Ψ can be performed, besides for a gaseous phase, only for a *dilute solution, i.e.* for a phase in which one kind of molecule far outnumbers all the others in the phase. We shall in future call this kind of molecule the *solvent* (cf. § 220), the other kinds the *dissolved substances*. If n_0 be the number of molecules of the solvent, n_1, n_2, n_3, . . . the number of molecules of the dissolved substances, then the solution may be considered dilute if n_0 be large in comparison with the sum of the numbers n_1, n_2, n_3. . . . The state of aggregation of the substance is of no importance, it may be solid, liquid, or gaseous.

§ 250. We shall now determine by the above method the energy U and the volume V of a dilute solution. The important simplification, to which this definition of a dilute solution leads, rests on the mathematical theorem, that a finite and continuous function of several variables with their differential coefficients which have very small values, is necessarily a *linear* function of these variables. This determines U and V as functions of n_0, n_1, n_2, ... Physically speaking, this means that the properties of a dilute solution, besides depending on the interactions between the molecules of the solvent, necessarily depend only on the interactions between the molecules of the solvent and the molecules of the dissolved substances, but not on the interactions of the dissolved substances among themselves, for these are small quantities of a higher order.

§ 251. The quotient $\dfrac{U}{n_0}$, *i.e.* the internal energy divided by the number of molecules of the solvent, remains unchanged if the numbers, n_0, n_1, n_2 ... be varied in the same proportion; for, by § 201, U is a homogeneous function of the number of molecules n_0, n_1, n_2, ..., of the first degree. $\dfrac{U}{n_0}$ is, therefore, a function of the ratios $\dfrac{n_1}{n_0}$, $\dfrac{n_2}{n_0}$, ..., and also a linear function, since these ratios are small, and the function is supposed to be differentiable. The function is, therefore, of the form

$$\frac{U}{n_0} = u_0 + u_1 \frac{n_1}{n_0} + u_2 \frac{n_2}{n_0} + \dots$$

where, u_0, u_1, u_2 are quantities depending, not on the number of molecules, but only on the temperature T, the pressure p, and the nature of the molecules. In fact, u_0 depends only on the nature of the solvent, since the energy reduces to $n_0 u_0$ when $n_1 = 0 = n_2 = \dots$, and u_1 only on the nature of the first dissolved substance and the solvent, and so on. u_0, therefore, corresponds to the interactions between the molecules of the solvent, u_1 to those between the solvent and the first dissolved substance, and so on. This contains a refutation of an objection, which is often raised against the modern theory of dilute solutions, that it treats dilute solutions simply as gases, and takes no account of the influence of the solvent.

§ 252. If the dilution is not sufficient to warrant the use of this very simple form of the function U, a more accurate relation may be obtained by regarding the coefficients u_1, $u_2 \ldots$ not as independent of the number of molecules but as functions of the ratios $\dfrac{n_1}{n_0}, \dfrac{n_2}{n_0}, \ldots$ Then certain new constants are obtained, which correspond to the interaction of the dissolved molecules one with another. This should be a possible way of constructing a rational thermodynamical theory of sulutions of any concentration.* In this way is found an explanation of the abnormal behaviour of solutions of strong electrolytes (cf. § 273).

§ **253.** We shall here keep to the simplest form, and write

$$\left. \begin{aligned} U &= n_0 u_0 + n_1 u_1 + n_2 u_2 + \ldots \\ V &= n_0 v_0 + n_1 v_1 + n_2 v_2 + \ldots \end{aligned} \right\} \quad . \quad . \quad (209)$$

and

How far these equations correspond to the facts may be determined by the inferences to which they lead. If we dilute the solution still further by adding one molecule of the solvent in the same state of aggregation as the solution, keeping meanwhile the temperature T and the pressure p constant, the corresponding change of volume and the heat of reaction may be calculated from the above equations. One molecule of the pure solvent, at the same temperature and pressure, has the volume v_0 and the energy u_0. After dilution, the volume of the solution becomes

$$V' = (n_0 + 1)\, v_0 + n_1 v_1 + n_2 v_2 + \ldots$$

and the energy

$$U' = (n_0 + 1)\, u_0 + n_1 u_1 + n_2 u_2 + \ldots$$

The increase of volume brought about by the dilution is therefore

$$V' - (V + v_0),$$

i.e. the increase of volume is zero. The heat absorbed is, by the first law (47),

$$U' - (U + u_0) + p\{V' - (V + v_0)\}.$$

This also vanishes.

* Cf. H. Jahn, *Zeitschr. f. phys. Chemie*, **41**, p. 257, 1902.

These inferences presuppose that the number of molecules of the dissolved substances remain unchanged, *i.e.* that no chemical changes (*e.g.* changes of the degree of dissociation) are produced by the dilution. If such were the case, the number of molecules of the dissolved substances would have values in the equations for U' and V' different from those in the equations for U and V, and therefore would not disappear on subtraction. We may therefore enunciate the following proposition : *Further dilution of a dilute solution, if no chemical changes accompany the process, produces neither an appreciable change of volume nor an appreciable heat of reaction ;* or, in other words, *any change of volume or any heat of reaction produced by further dilution of a dilute solution is due to chemical transformations among the molecules of the dissolved substances* (cf. § 97).

§ **254.** We now turn to the calculation of the entropy Φ of a dilute solution. If the number of molecules n_0, n_1, n_2, ... be constant, we have, by (60),

$$d\Phi = \frac{dU + pdV}{T},$$

and, by (209),

$$d\Phi = n_0\frac{du_0 + pdv_0}{T} + n_1\frac{du_1 + pdv_1}{T} + n_2\frac{du_2 + pdv_2}{T} + \cdots$$

Since u and v are functions of T and p only, and not of n, each of the coefficients of n_0, n_1, n_2, ..., must be a perfect differential, *i.e.* there must be certain functions ϕ, depending only on T and p, such that

$$\left.\begin{aligned}d\phi_0 &= \frac{du_0 + pdv_0}{T}\\[4pt]d\phi_1 &= \frac{du_1 + pdv_1}{T}\\[4pt]d\phi_2 &= \frac{du_2 + pdv_2}{T}\end{aligned}\right\} \quad \cdots \cdots \quad (210)$$

We have, then,

$$\Phi = n_0\phi_0 + n_1\phi_1 + n_2\phi_2 + \ldots + C, \quad . \quad . \quad (211)$$

where the integration constant C cannot depend on T and p, but may be a function of the number of molecules. C may be determined as a function of $n_0, n_1, n_2, \ldots$ for a particular temperature and pressure, and this will be the general expression for C at any temperature and pressure.

We shall now determine C as a function of n taking the particular case of high temperature and small pressure. By increasing the temperature and diminishing the pressure, the solution, whatever may have been its original state of aggregation, will pass completely into the gaseous state. Chemical changes and changes in the state of aggregation will certainly take place at the same time, *i.e.* the number of molecules n will change. Partial evaporation, etc., will set in. Only such states as lie very close to stable states are realisable in nature. We shall, however, assume that the process takes place in such a way as to leave the number of the different kinds of molecules unaltered, and that the whole system consists only of one phase, for only in this case does C retain its value. This supposition is permissible, since the number of molecules n, together with T and p, are the independent variables of the system. Such a process is possible only in the ideal sense, since it passes through unstable states. There is, however, no objection to its use for our present purpose, since the above expression for Φ holds not only for stable states of equilibrium, but for all states characterized by quite arbitrary values of T, p, n_0, n_1, n_2, . . . Stable equilibrium is a special case, satisfying a further condition to be established below.

At a sufficiently high temperature, and a sufficiently low pressure, any gaseous system possesses so small a density, that it may be regarded as a mixture of perfect gases (§ 21, and § 43). We have, therefore, by (194), bearing in mind that here the first kind of molecule is denoted by the suffix 0,

$$\Phi = n_0(C_{p_0} \log T - R \log p + k_0)$$
$$+ n_1(C_{p_1}\log T - R \log p + k_1) + \ldots + C. \quad (212)$$

The constant C is independent of T and p, and has the value given in (198). On comparing this with (211), it is seen that the expression for Φ can pass from (211) into (212) by mere change of temperature and pressure, only if the constant C is the same in both expressions, *i.e.* if, by (198),

$$C = - R(n_0 \log c_0 + n_1 \log c_1 + \ldots)$$

Here the concentrations are

$$c_0 = \frac{n_0}{n_0 + n_1 + n_2 \ldots}; \quad c_1 = \frac{n_1}{n_0 + n_1 + n_2 \ldots}.$$

By (211), the entropy of a dilute solution at a definite temperature and pressure becomes

$$\Phi = n_0(\phi_0 - R \log c_0) + n_1(\phi_1 - R \log c_1) + \ldots \quad (213)$$

If we put, for shortness, the quantities which depend only on T and p, and not on n the number of molecules

$$\left. \begin{aligned} \phi_0 - \frac{u_0 + pv_0}{T} &= \psi_0 \\ \phi_1 - \frac{u_1 + pv_1}{T} &= \psi_1 \\ \phi_2 = \frac{u_2 + pv_2}{T} &= \psi_2 \end{aligned} \right\} \quad \ldots \quad (214)$$

we have, finally, from (75), (213) and (209),

$$\Psi = n_0(\psi_0 - R \log c_0) + n_1(\psi_1 \, R \log c_1) \\ + n_2(\psi_2 - R \log c_2) + \ldots \quad (215)$$

This equation determines the thermodynamical properties of a dilute solution. The characteristic function Ψ has a form quite similar to that for a gas mixture in (199a). The only difference lies in the fact that the quantities ψ here depend on the nature of the solvent.

§ 255. We may now proceed to establish the conditions of equilibrium of a system consisting of several phases. As hitherto, the different kinds of molecules in the phase will be denoted by suffixes, and the different phases by dashes.

For the sake of simplicity the first phase will be left without a dash. The entire system is then represented by

$$n_0\, m_0,\, n_1\, m_1,\, n_2\, m_2,\, \ldots \mid n_0'\, m_0',\, n_1'\, m_1',\, n_2'\, m_2',\, \ldots$$
$$\mid n_0''\, m_0'',\, n_1''\, m_1'',\, n_2''\, m_2'',\, \ldots \mid \quad . \quad . \quad (216)$$

The number of molecules is denoted by n, and the molecular weights by m, and the individual phases are separated by vertical lines. In the general formula we signify the summation over the different kinds of molecules of one and the same phase by writing the individual terms of the summation; the summation over the different phases, on the other hand,

by the symbol $\sum$.

To enable us to apply the derived formulæ, we shall assume that each phase is either a mixture of perfect gases or a dilute solution. The latter designation will be applied to phases containing only one kind of molecule, *e.g.* a chemically homogeneous solid precipitate from an aqueous solution. One kind of molecule represents the special case of a dilute solution in which the concentrations of all the dissolved substances are zero.

§ 256. Suppose now that an isothermal isobaric change be possible, corresponding to a simultaneous variation $\delta n_0,\, \delta n_1,\, \delta n_2,\, \ldots \delta n_0',\, \delta n_1',\, \delta n_2',\, \ldots$ of the number of molecules $n_0,\, n_1,\, n_2,\, \ldots n_0',\, n_1',\, n_2',\, \ldots$; then, by (79), this change will not take place, if at constant temperature and pressure

$$\delta \Psi = 0$$

or, by (215), if

$$\sum (\psi_0 - R \log c_0)\delta n_0 + (\psi_1 - R \log c_1)\delta n_1 + (\psi_2 - R \log c_2)\delta n_2$$
$$+ \ldots + \sum n_0 \delta(\psi_0 - R \log c_0) + n_1 \delta(\psi_1 - R \log c_1)$$
$$+ n_2 \delta(\psi_2 - R \log c_2) + \ldots = 0.$$

The summation $\sum$ extends over all the phases of the system. The second series is identically equal to zero for the same reason as was given in connection with equation (200). If we again introduce the simple integral ratios

$$\delta n_0 : \delta n_1 : \delta n_2 : \ldots : \delta n_0' : \delta n_1' : \delta n_2' : \ldots$$
$$= \nu_0 : \nu_1 : \nu_2 : \ldots : \nu_0' : \nu_1' : \nu_2' : \ldots \quad (217)$$

then the equation of equilibrium becomes

$$\sum (\psi_0 - \mathrm{R} \log c_0)\nu_0 + (\psi_1 - \mathrm{R} \log c_1)\nu_1$$
$$+ (\psi_2 - \mathrm{R} \log c_2)\nu_2 + \ldots = 0$$

or

$$\sum \nu_0 \log c_0 + \nu_1 \log c_1 + \nu_2 \log c_2 + \ldots$$
$$= \frac{1}{\mathrm{R}} \sum \nu_0 \psi_0 + \nu_1 \psi_1 + \ldots = \log \mathrm{K}. \quad (218)$$

K like ψ_0, ψ_1, ψ_2, is independent of the number of molecules n.

§ **257.** The definition of K gives its functional relation to T and p.

$$\frac{\partial \log \mathrm{K}}{\partial \mathrm{T}} = \frac{1}{\mathrm{R}} \sum \nu_0 \frac{\partial \psi_0}{\partial \mathrm{T}} + \nu_1 \frac{\partial \psi_1}{\partial \mathrm{T}} + \nu_2 \frac{\partial \psi_2}{\partial \mathrm{T}} + \ldots$$

$$\frac{\partial \log \mathrm{K}}{\partial p} = \frac{1}{\mathrm{R}} \sum \nu_0 \frac{\partial \psi_0}{\partial p} + \nu_1 \frac{\partial \psi_1}{\partial p} + \nu_2 \frac{\partial \psi_2}{\partial p} + \ldots$$

Now, by (214), we have for an infinitely small change of T and p

$$d\psi_0 = d\phi_0 - \frac{du_0 + p\,dv_0 + v_0\,dp}{\mathrm{T}} + \frac{u_0 + pv_0}{\mathrm{T}^2} d\mathrm{T}$$

and therefore, by (210),

$$d\psi_0 = \frac{u_0 + pv_0}{\mathrm{T}^2} d\mathrm{T} - \frac{v_0\,dp}{\mathrm{T}}.$$

From this it follows that

$$\frac{\partial \psi_0}{\partial \mathrm{T}} = \frac{u_0 + pv_0}{\mathrm{T}^2}; \quad \frac{\partial \psi_0}{\partial p} = -\frac{v_0}{\mathrm{T}}.$$

Similarly

$$\frac{\partial \psi_1}{\partial \mathrm{T}} = \frac{u_1 + pv_1}{\mathrm{T}^2}; \quad \frac{\partial \psi_1}{\partial p} = -\frac{v_1}{\mathrm{T}}.$$

Hence

$$\frac{\partial \log K}{\partial T} = \frac{1}{RT^2}\sum(\nu_0 u_0 + \nu_1 u_1 + \ldots) + p(\nu_0 v_0 + \nu_1 v_1 + \ldots),$$

$$\frac{\partial \log K}{\partial p} = -\frac{1}{RT}\sum\nu_0 v_0 + \nu_1 v_1 + \ldots$$

Denoting by v the increase of volume of the system, and by L the heat absorbed, when the change corresponding to (217) takes place at constant temperature and pressure, then, by (209),

$$v = \sum \nu_0 v_0 + \nu_1 v_1 + \nu_2 v_2 + \ldots$$

and, by the first law of thermodynamics,

$$L = \sum(\nu_0 u_0 + \nu_1 u_1 + \ldots) + p(\nu_0 v_0 + \nu_1 v_1 + \ldots);$$

therefore
$$\frac{\partial \log K}{\partial T} = \frac{L}{RT^2} \quad \cdots \quad (219)$$

and
$$\frac{\partial \log K}{\partial p} = -\frac{v}{RT} \quad \cdots \quad (220)$$

The influence of the temperature on K, and therewith on the condition of equilibrium towards a certain chemical reaction, is controlled by the heat of that reaction, and the influence of the pressure is controlled by the corresponding change of volume of the system. If the reaction take place without the absorption or evolution of heat, the temperature has no influence on the equilibrium. If it produce no change of volume the pressure has no influence (cf. § 211).

The elimination of K from these two equations gives a general relation between the heat of reaction L and the change of volume, v;

$$\frac{\partial L}{\partial p} = v - T\frac{\partial v}{\partial T},$$

in agreement with the general equation (79g). The former equations (205) and (206) are particular cases of (219) and

(220), as may be seen by substituting for log K the special
value given in (203a),

$$\log K = \log A - \frac{B}{T} + C \log T - \nu \log p.$$

§ 258. By means of equation (218) a condition of equilibrium may be established for each possible change in a given
system subject to chemical change. Of course, K will have
a different value in each case. This corresponds to the
requirements of Gibbs's phase rule, which is general in its
application (§ 204). The number of the different kinds of
molecules in the system must be distinguished from the number
of the independent constituents (§ 198). Only the latter
determines the number and nature of the phases; while the
number of the different kinds of molecules plays no part
whatever in the application of the phase rule. If another
kind of molecule be introduced the number of the variables
increases, to be sure, but so does the number of the possible
reactions, and therewith, the number of the conditions of
equilibrium by the same amount, so that the number of
independent variables is quite independent thereof.

§ 259. Equation (218) shows further that, generally
speaking, all kinds of molecules possible in the system
will be present in finite numbers in every phase; for instance,
molecules of water must occur in any precipitate from an
aqueous solution. Even solid bodies in contact must partially
dissolve in one another, if sufficient time be given. The
quantity K, which determines the equilibrium, possesses,
according to the definition (218), a definite, finite value for
each possible chemical change, and none of the concentrations c can, therefore, vanish so long as the temperature
and the pressure remain finite. This principle, based entirely
on thermodynamical considerations, has already served to
explain certain facts, e.g. the impossibility of removing the
last traces of impurity from gases, liquids, and even solids. It
also follows from it that absolutely semipermeable membranes
are non-existent, for the substance of any membrane would,

in time, become saturated with the molecules of all the various kinds of substances in contact with one side of it, and thus give up each kind of substance to the other side (cf. § 229).

On the other hand, this view greatly complicates the calculation of the thermodynamical properties of a solution, since, in order to make no mistake, it is necessary to assume from the start the existence in every phase of all kinds of molecules possible from the given constituents. We must not neglect any kind of molecule until we have ascertained by a particular experiment that its quantity is inappreciable. Many cases of apparent discrepancy between theory and experiment may probably be explained in this way.

§ **259A.** All the foregoing propositions refer, of course, to finite values of temperature and pressure. If, however, the temperature T approaches the absolute zero, a glance at equation (219) shows that the heat of the reaction, L, remains finite, while log K becomes infinite, positive or negative, according to the direction of the reaction shown by the sign of the numbers ν. From this it follows, that at the absolute zero of temperature the reaction goes completely in one direction or the other, so that at last the concentrations of the molecules, which are being decomposed by the reaction, become zero. This result agrees with the general conclusion of § 144, that at low temperatures the reactions proceed with an evolution of heat. It still further defines that conclusion since with dilute solutions it is, in general, only possible to draw conclusions from the total energy and not directly from the free energy. We may therefore annunciate the following general law. *If the temperature decreases indefinitely, all molecules which are capable of transformations which evolve heat, vanish from a solution, which is in thermodynamical equilibrium.*

We shall now discuss some of the most important particular cases. They have been arranged, in the first place, according to the number of the independent constituents of the system; in the second, according to the number of the phases.

§ 260. One Independent Constituent in One Phase.

—According to the phase rule, the nature of the phase depends on two variables, *e.g.* on the temperature and the pressure. The phase may contain any number of different kinds of molecules. Water, for instance, will contain simple, double, and multiple H_2O molecules, molecules of hydrogen and oxygen, H_2 and O_2, also H_2O_2 molecules and electrically charged ions $\overset{+}{H}$, $\overset{-}{HO}$, and $\overset{--}{O}$, etc., in finite quantities. The electrical charges of the ions do not play any important part in thermodynamics, so long as there is no direct conflict between the electrical and the thermodynamical forces. This happens when and only when the thermodynamical conditions of equilibrium call for such a distribution of the ions in the different phases of the system as would lead, on account of the constant charges of the ions, to free electricity in any phase. The electrical forces strongly oppose such a distribution, and the resulting deviation from the pure thermodynamical equilibrium is, however, compensated by differences of potential between the phases. A general view of these electromolecular phenomena may be got by generalizing the expressions for the entropy and the energy of the system by the addition of electrical terms. We shall restrict our discussion to states which do not involve electrical phenomena, and need not consider the charges of the ions. We may treat the ions like other molecules.

In the case mentioned above, then, the concentrations of all kinds of molecules are determined by T and p. The calculation of the concentrations has succeeded so far only in the case of the $\overset{+}{H}$ and $\overset{-}{OH}$ ions (the number of the $\overset{--}{O}$ ions is negligible), in fact, among other methods, by the measurement of the electrical conductivity of the solution, which depends only on the ions. Kohlrausch and Heydweiller found the degree of dissociation of water, *i.e.* the ratio of the mass of water split into $\overset{+}{H}$ and $\overset{-}{OH}$ ions to the total mass of water to be, at 18° C.,

$$14 \cdot 3 \times 10^{-10}.$$

This number represents the ratio of the number of dissociated molecules to the total number of molecules. We may determine by thermodynamics the temperature coefficient of the dissociation.

The condition of equilibrium will now be established. The system is, by (216),

$$n_0 \; H_2O, \; n_1 \; \overset{+}{H}, \; n_2 \; \overset{-}{OH}.$$

Let the total number of molecules be

$$n = n_0 + n_1 + n_2,$$

the concentrations are, therefore,

$$c_0 = \frac{n_0}{n}, \; c_1 = \frac{n_1}{n}, \; c_2 = \frac{n_2}{n}.$$

The chemical reaction in question,

$$\nu_0 : \nu_1 : \nu_2 = \delta n_0 : \delta n_1 : \delta n_2,$$

consists in the dissociation of one H_2O molecule into $\overset{+}{H}$ and $\overset{-}{OH}$.

$$\nu_0 = -1; \; \nu_1 = 1; \; \nu_2 = 1;$$

and therefore, by (218), in the state of equilibrium

$$- \log c_0 + \log c_1 + \log c_2 = K,$$

or, since $c_1 = c_2$, and $c_0 = 1$ nearly,

$$2 \log c_1 = \log K.$$

This gives, by (219), the relation between the concentration and the temperature :

$$2 \frac{\partial \log c_1}{\partial T} = \frac{1}{R} \cdot \frac{L}{T^2} \quad \cdot \quad \cdot \quad \cdot \quad \cdot \quad (221)$$

According to Arrhenius, L, the heat necessary for the dissociation of one molecule of H_2O into $\overset{+}{H}$ and $\overset{-}{OH}$, is equal to the heat of neutralization of a strong monobasic acid and base in dilute aqueous solution.

According to the notation of § 97,

$$L = (\overset{+}{H}, \overset{-}{Cl}, aq) + (\overset{+}{Na}, \overset{-}{OH}, aq) - (\overset{+}{Na}, \overset{-}{Cl}, aq).$$

This gives according to the recent measurements of Wörmann :

$$L = 27857 - 48{\cdot}5 \ T \ cal.$$

From equation (221) it follows that

$$\frac{\partial \log c_1}{\partial T} = \frac{1}{2 \times 1{\cdot}985}\left(\frac{27857}{T^2} - \frac{48{\cdot}5}{T}\right),$$

and, by integration,

$$\log_{10} c_1 = -\frac{3047{\cdot}3}{T} - 12{\cdot}125 \log_{10} T + \text{const.}$$

This equation, which shows how the degree of dissociation depends on the temperature, agrees well with the measurements of the electrical conductivity of pure water at different temperatures by Kohlrausch and Heydweiller, Noyes, and Lunden.

At the absolute zero of temperature, the dissociation vanishes completely, in agreement with the general law deduced in § 259A.

§ 261. One Independent Constituent in Two Phases.
The system consists of two phases let us say of a liquid and a gaseous or solid phase. By (216), the system is represented by

$$n_0 \, m_0 \mid n_0{}' \, m_0{}'.$$

Each phase contains only one kind of molecule. The molecules in both phases need not however be the same. The liquid molecules may be any multiple of the gaseous molecules. If now a liquid molecule evaporate or solidifies, then, in our notation,

$$\nu_0 = -1, \ \nu_0{}' = \frac{m_0}{m_0{}'},$$

$$c_0 = \frac{n_0}{n_0} = 1, \ c_0{}' = \frac{n_0{}'}{n_0{}'} = 1$$

and, according to the equilibrium condition (218),

$$0 = \log K = - \psi_0 + \frac{m_0}{m_0'}\psi_0' \quad . \quad . \quad (221a)$$

Since K depends only on the temperature and the pressure, this equation gives a definite relation between temperature and pressure : viz., the law which states how the pressure of evaporation or fusion depends on the temperature, and vice versa. The meaning of this law can be obtained by considering how K depends on p and T. On differentiating the last equation, we get,

$$\frac{\partial \log K}{\partial T}dT + \frac{\partial \log K}{\partial p}dp = 0,$$

or, by (219) and (220),

$$\frac{L}{T^2}dT - \frac{v}{T}dp = 0.$$

If v_0 and v_0' be the molecular volumes of the two phases, then the change of volume of the system for the transformation under consideration is

$$v = \frac{m_0}{m_0'}v_0' - v_0,$$

hence,

$$L = T\Big(\frac{m_0}{m_0'}v_0' - v_0\Big)\frac{dp}{dT},$$

or, per unit mass,

$$\frac{L}{m_0} = T\Big(\frac{v_0'}{m_0'} - \frac{v_0}{m_0}\Big)\frac{dp}{dT},$$

the well-known Carnot-Clapeyron equation (111).

(For further applications see Chapter II.)

§ 262. **Two Independent Constituents in One Phase.**—(A substance dissolved in a homogeneous solvent.) According to the phase rule, one other variable besides the pressure and the temperature is arbitrary, *e.g.* the number of the molecules dissolved in 1 litre of the solution, a quantity which may be directly measured. The values of these three variables determine the concentrations of all kinds of molecules, whether they have their origin in dissociation, association, formation of hydrates, or hydrolysis of the dissolved molecules. Let us consider the simple case of a binary electrolyte, *e.g.* acetic acid in water. The system is represented by

$$n_0 \; H_2O, \; n_1 \; CH_3.COOH, \; n_2 \; \overset{+}{H}, \; n_3 \; \overset{-}{CH_3}.COO.$$

The total number of molecules,

$$n = n_0 + n_1 + n_2 + n_3,$$

is only slightly greater than n_0. The concentrations are

$$c_0 = \frac{n_0}{n}; \; c_1 = \frac{n_1}{n}; \; c_2 = \frac{n_2}{n}; \; c_3 = \frac{n_3}{n}.$$

The only reaction

$$\nu_0 : \nu_1 : \nu_2 : \nu_3 = \delta n_0 : \delta n_1 : \delta n_2 : \delta n_3,$$

which need be considered consists in the dissociation of one molecule of $CH_3.COOH$ into its two ions.

$$\nu_0 = 0; \; \nu_1 = -1; \; \nu_2 = 1; \; \nu_3 = 1.$$

Therefore, in equilibrium,

$$- \log c_1 + \log c_2 + \log c_3 = \log K;$$

or, since

$$c_2 = c_3,$$

$$\frac{c_2^2}{c_1} = K \quad . \quad . \quad . \quad . \quad (222)$$

Now, we may regard the sum

$$c_1 + c_2 = c$$

as known, since the total number $(n_1 + n_2)$ of the undissociated and the dissociated molecules of the acid, and the total number of water molecules, which may be put $= n$, are measured directly. Hence c_1 and c_2 may be calculated from the last two equations.

$$\frac{c_1}{c} = \frac{n_1}{n_1 + n_2} = 1 - \frac{K}{2c}\left(\sqrt{1 + \frac{4c}{K}} - 1\right);$$

$$\frac{c_2}{c} = \frac{n_2}{n_1 + n_2} = \frac{K}{2c}\left(\sqrt{1 + \frac{4c}{K}} - 1\right).$$

With increasing dilution (decreasing c), the ratio $\frac{c_2}{c}$ increases in a definite manner approaching the value 1, *i.e.* complete dissociation. This also gives for the electrical conductivity of a solution of given concentration Ostwald's so-called *law of dilution of binary electrolytes*, which has been experimentally verified in numerous cases.

By considering the heat of dissociation we can, as in § 260, by equation (219) find how K depends on the temperature, and conversely from the change of the dissociation with temperature we can calculate the heat of dissociation, as was first shown by Arrhenius.*

The observed deviations from Ostwald's dilution law in the case of strong electrolytes may be explained by the fact that the electrical conductivity of a solution is not always a measure of the state of the dissociation. According to J. C. Ghosh.† the salt molecules in dilute solutions of strong electrolytes, like KCl and NaCl, are as good as completely dissociated into their ions, but only a certain fraction of the ions contribute to the electrical conduction. The slower moving ions are held back by the attractions of the neighbouring oppositely charged ions. This fraction increases with increasing dilution on account of the decrease of the influence of these attractions. Accordingly, the conductivity increases with dilution according to a law which cannot be deduced here, since in electrical conduction a state of equilibrium does not obtain. This law deviates in essential points from Ostwald's dilution law, as also from the laws of the lowering of the freezing point and of the osmotic pressure of strong electrolytes, deduced in § 273.

§ 263. Usually, however, in a solution, not one, but a large number of reactions will be possible. Accordingly, the complete system contains many kinds of molecules. As another example, we shall discuss the case of an electrolyte

---

* The variation of K with pressure, as given by equation (220) was experimentally tested and verified by Fanjung.

† Jnanendra Chandra Ghosh, *Trans. Chem. Soc.*, **113**, p. 449, 1918.

capable of splitting into ions in several ways, viz. an aqueous solution of sulphuric acid. The system is represented by

$$n_0 \ H_2O, \ n_1 \ H_2SO_4, \ n_2 \ \overset{+}{H}, \ n_3 \ \overset{-}{HSO_4}, \ n_4 \ \overset{--}{SO_4}.$$

The total number of molecules is

$$n = n_0 + n_1 + n_2 + n_3 + n_4 \ \text{(nearly equal to } n_0\text{)}.$$

The concentrations are

$$c_0 = \frac{n_0}{n}; \ c_1 = \frac{n_1}{n}; \ c_2 = \frac{n_2}{n}; \ c_3 = \frac{n_3}{n}; \ c_4 = \frac{n_4}{n}.$$

Here two different kinds of reactions

$$\nu_0 : \nu_1 : \nu_2 : \nu_3 : \nu_4 = \delta n_0 : \delta n_1 : \delta n_2 : \delta n_3 : \delta n_4$$

must be considered; first, the dissociation of one molecule of H_2SO_4 into $\overset{+}{H}$ and $\overset{-}{HSO_4}$:

$$\nu_0 = 0; \ \nu_1 = -1; \ \nu_2 = 1; \ \nu_3 = 1; \ \nu_4 = 0;$$

second, the dissociation of the ion $\overset{-}{HSO_4}$ into $\overset{+}{H}$ and $\overset{--}{SO_4}$

$$\nu_0 = 0; \ \nu_1 = 0; \ \nu_2 = 1; \ \nu_3 = -1; \ \nu_4 = 1.$$

Hence, by (218), there are two conditions of equilibrium :

$$- \log c_1 + \log c_2 + \log c_3 = \log K$$

and
$$\log c_2 - \log c_3 + \log c_4 = \log K';$$

or
$$\frac{c_2 \, c_3}{c_1} = K$$

and
$$\frac{c_2 \, c_4}{c_3} = K'.$$

This further condition must be added, viz. that the total number of SO_4 radicals $(n_1 + n_3 + n_4)$ must be equal to half the number of H atoms $(2n_1 + n_2 + n_3)$; otherwise the system would contain more than two independent constituents. This condition is

$$2c_4 + c_3 = c_2.$$

Finally, the quantity of sulphuric acid in the solution is supposed to be given :

$$c_1 + c_3 + c_4 = c.$$

The last four equations determine c_1, c_2, c_3, c_4, and hence the state of equilibrium is found.*

For a more accurate determination it would be necessary to consider still other kinds of molecules. Every one of these introduces a new variable, but also a new possible reaction, and therefore a new condition of equilibrium, so that the state of equilibrium remains uniquely determined.

§ **264. Two Independent Constituents in Two Phases.**—The state of equilibrium, by the phase rule, depends on two variables, *e.g.* temperature and pressure. The wide range of cases in point makes a subdivision desirable, according as only one phase contains both constituents in appreciable quantity, or both phases contains both constituents.

Let us first take the simpler case, where one (first) phase contains both constituents, and the other (second) phase contains only one single constituent. Strictly speaking this never occurs (by 259), but in many cases it is a sufficient approximation to the actual facts. The application of the general condition of equilibrium (218) to this case leads to different laws, according as the constituent in the second phase plays the part of dissolved substance or solvent (§ 249) in the first phase. We shall therefore divide this case into two further subdivisions.

§ **265. The Pure Substance in the Second Phase forms the Dissolved Body in the First.**—An example of this is the absorption of a gas, *e.g.* carbon dioxide in a liquid of comparatively small vapour pressure; *e.g.* water at not too high a temperature.

The system is represented by

$$n\ H_2O,\ n_1\ CO_2 \mid n_0'\ CO_2.$$

* See the calculations of J. B. Goebel, *Zeitschr. f. physikal. Chemie,* **71**, S. 652, 1910.

We assume that the gas molecule CO_2 is identical with the dissolved molecule CO_2. The general case will be further dealt with in § 274. The concentrations of the different kinds of molecule of the system in the two phases are

$$c_0 = \frac{n_0}{n_0 + n_1}; \quad c_1 = \frac{n_1}{n_0 + n_1}; \quad c_0' = \frac{n_0'}{n_0'} = 1.$$

The reaction

$$v_0 : v_1 : v_0' = \delta n_0 : \delta n_1 : \delta n_0'$$

consists in the evaporation of one molecule of carbon dioxide from the solution, therefore,

$$v_0 = 0, \, v_1 = -1, \, v_0' = 1.$$

The condition of equilibrium

$$v_0 \log c_0 + v_1 \log c_1 + v_0' \log c_0' = \log K,$$

is, therefore,

$$- \log c_1 = \log K, \quad . \quad . \quad . \quad (223)$$

or, at a given temperature and pressure (for these determine K), c_1 the concentration of the gas in the solution is determined. The change of concentration with pressure and temperature is found by substituting (223) in (219) and (220) :

$$\frac{\partial \log c_1}{\partial p} = \frac{1}{R} \cdot \frac{v}{T} \cdot \quad . \quad . \quad . \quad . \quad (224)$$

$$\frac{\partial \log c_1}{\partial T} = -\frac{1}{R} \cdot \frac{L}{T^2} \cdot \quad . \quad . \quad . \quad (225)$$

v is the increase of volume of the system, L the heat absorbed during isothermal and isobaric evaporation of one gram molecule of CO_2. Since v represents nearly the volume of one gram molecule of carbon dioxide gas, we may, by (16), put

$$v = \frac{RT}{p},$$

and equation (224) gives

$$\frac{\partial \log c_1}{\partial p} = \frac{1}{p}.$$

On integrating, we have

$$\log c_1 = \log p + \text{const.}$$

or $$c_1 = Cp \quad . \quad . \quad . \quad . \quad . \quad . \quad (226)$$

i.e. the concentration of the dissolved gas is proportional to the pressure of the free gas on the solution (Henry's law). The factor C, which is a measure of the solubility of the gas, still depends on the temperature, since (225) and (226) give

$$\frac{\partial \log C}{\partial T} = -\frac{1}{R} \cdot \frac{L}{T^2}.$$

If, therefore, heat is absorbed during the evaporation of the gas from the solution, L is positive, and the solubility decreases with increase of temperature. Conversely, from the variation of C with temperature, the heat of absorption may be calculated;

$$L = -\frac{RT^2}{C} \cdot \frac{\partial C}{\partial T}.$$

According to the experiments of Naccari and Pagliani, the solubility of carbon dioxide in water at 20° (T = 293), (expressed in a unit which need not be discussed here), is 0·8928, its temperature coefficient − 0·02483; therefore, by (34),

$$L = \frac{1·985 \times 293^2 \times 0·02483}{0·8928} = 4700 \text{ cal.}$$

Thomsen found the heat of absorption of one gram molecule of carbon dioxide to be 5880 cal. The error (according to Nernst) lies mainly in the determination of the coefficient of solubility. Of this total heat an amount

$$RT = 1·985 \times 293 = 590 \text{ cal.}$$

corresponds, by (48), to external work.

§ **266.** A further example is the saturation of a liquid with an almost insoluble salt; *e.g.* succinic acid in water. The system is represented by

$$n_0 \, H_2O, \, n_1 \begin{vmatrix} CH_2 - COOH \\ | \\ CH_2 - COOH, \end{vmatrix} \, n_0' \begin{vmatrix} CH_2 - COOH \\ | \\ CH_2 - COOH, \end{vmatrix}$$

if the slight dissociation of the acid in water be neglected. The calculation of the condition of equilibrium gives, as in § 223,

$$- \log c_1 = \log K,$$

c_1 is determined by temperature and pressure. Further, by (219),

$$L = - RT^2 \frac{\partial \log c_1}{\partial T} \quad . \quad . \quad . \quad (227)$$

Van't Hoff was the first to calculate L by means of this equation from the solubility of succinic acid at 0° C. (2·88) and at 8·5° C. (4·22)

$$\frac{\partial \log c_1}{\partial T} = \frac{\log_e 4 \cdot 22 - \log_e 2 \cdot 88}{8 \cdot 5} = 0 \cdot 04494.$$

This gives, for $T = 273$, $L = - 1·985 \times 273^2 \times 0·4494 = - 6600$ cals.; *i.e.* on the precipitation of one gram molecule of the solid from the solution, 6600 cals. are given out. Berthelot found the heat of solution to be 6700 cals.

If L be regarded as independent of the temperature, which is permissible in many cases as a first approximation, the equation (227) may be integrated with respect to T, giving

$$\log c_1 = \frac{L}{RT} + \text{const.}$$

§ 267. The relation (227) becomes inapplicable if the salt in solution undergoes an appreciable chemical transformation, *e.g.* dissociation. For then, besides the ordinary molecules of the salt, the products of the dissociation are present in the solution; for example, in the system of water and silver acetate,

$$n_0 H_2O, \; n_1 CH_3COOAg, \; n_2 \overset{+}{Ag}, \; n_3 CH_3 . \overset{-}{COO} \mid n_0' CH_3COOAg.$$

The total number of molecules in the solution:

$$n = n_0 + n_1 + n_2 + n_3 \text{ (nearly} = n_0).$$

The concentrations of the different molecules in both phases
are

$$c_0 = \frac{n_0}{n}; \ c_1 = \frac{n_1}{n}; \ c_2 = \frac{n_2}{n}; \ c_3 = \frac{n_3}{n}; \ c_0' = \frac{n_0'}{n_0'} = 1.$$

The reactions,

$$\nu_0 : \nu_1 : \nu_2 : \nu_3 : \nu_0' = \delta n_0 : \delta n_1 : \delta n_2 : \delta n_3 : \delta n_0',$$

are :

(1) The precipitation of a molecule of the salt from the
solution :

$$\nu_0 = 0, \nu_1 = -1, \nu_2 = 0, \nu_3 = 0, \nu_0' = 1.$$

(2) The dissociation of a molecule of silver acetate :

$$\nu_0 = 0, \ \nu_1 = -1, \nu_2 = 1, \nu_3 = 1, \nu_0' = 0.$$

Accordingly, the two conditions of equilibrium are :

(1) $- \log c_1 = \log K$

(2) $- \log c_1 + \log c_2 + \log c_3 = \log K';$

or, since $c_2 = c_3$,

$$\frac{c_2^2}{c_1} = K'.$$

At given temperature and pressure, therefore, there is in
the saturated solution of a salt a definite number of undissociated molecules; and the concentration (c_2) of the dissociated molecules may be derived from that of the undissociated
(c_1) by the law of dissociation of an electrolyte, as given in
(222).

Now, since by measuring the solubility the value of
$c_1 + c_2$, and by measuring the electrical conductivity the
value of c_2, may be found, the quantities K and K' can be
calculated for any temperature. Their dependence on temperature, by (219), serves as a measure of the heat of the
precipitation of an undissociated molecule from the solution,
and of the dissociation of a dissolved molecule. Van't Hoff
has thus given a method of calculating the actual heat of
solution of a salt, from measurements of the solubility of the

salt and of the conductivity of saturated solutions at different temperatures; *i.e.* the heat of the reaction when one gram molecule of the solid salt is dissolved, and the fraction $\dfrac{dc_2}{dc_1 + dc_2}$ is dissociated into its ions, as is actually the case in the process of solution.

§ 268. **The Pure Substance occurring in the Second Phase forms the Solvent in the First Phase.**—This case is realized when the pure solvent in any state of aggregation is separated out from a solution of another state of aggregation, *e.g.* by freezing, evaporation, fusion, and sublimation. The type of such a system is

$$n_0 \; m_0, \; n_1 \; m_1, \; n_2 \; m_2, \; n_3 \; m_3, \; \ldots \mid n_0' \; m_0'.$$

The question whether the solvent has the same molecular weight in both phases, or not, is left open. The total number of molecules in the solution is

$$n = n_0 + n_1 + n_2 + n_3 + \ldots \text{(nearly} = n_0).$$

The concentrations are,

$$c_0 = \frac{n_0}{n}; \; c_1 = \frac{n_1}{n}; \; c_2 = \frac{n_2}{n}; \; \ldots \; c_0' = \frac{n_0'}{n_0'} = 1.$$

A possible transformation,

$$\nu_0 : \nu_1 \ldots : \nu_0' = \delta n_0 = \delta n_1 \ldots : \delta n_0',$$

is the passage of a molecule of the solvent from the first phase to the second phase, *i.e.*

$$\nu_0 = -1; \; \nu_1 = 0; \; \nu_2 = 0; \; \ldots \nu_0' = \frac{m_0}{m_0'}. \quad (228)$$

Equilibrium demands, by (218), that

$$-\log c_0 + \frac{m_0}{m_0'} \log c_0' = \log \mathrm{K},$$

and, therefore, on substituting the above values of c_0 and c_0'

$$\log \frac{n}{n_0} = \log \mathrm{K}.$$

But
$$\frac{n}{n_0} = 1 + \frac{n_1 + n_2 + n_3 + \cdots}{n_0},$$

and, therefore, since the fraction on the right is very small,

$$\frac{n_1 + n_2 + n_3 + \cdots}{n_0} = \log K. \quad . \quad . \quad (229)$$

By the general definition (218), we have

$$\log K = \frac{1}{R}(\nu_0 \psi_0 + \nu_1 \psi_1 + \nu_2 \psi_2 + \cdots + \nu_0' \psi_0'),$$

and, therefore, on substituting the values of ν from (228),

$$\frac{n_1 + n_2 + n_3 + \cdots}{n_0} = \frac{1}{R}\left(\frac{m_0}{m_0'}\psi_0' - \psi_0\right) = \log K \quad . \quad (230)$$

This expression shows that $\log K$ also has a small value.

Suppose for the moment that $\log K = 0$, *i.e.* that the pure solvent takes the place of the solution

$$n_1 + n_2 + \cdots = 0,$$

then, by (230),

$$\frac{\psi_0}{m_0} = \frac{\psi_0'}{m_0'}.$$

This equation asserts a definite relation which the temperature and the pressure must fulfil, in order that the two states of aggregation of the pure solvent may exist in contact. The pressure (vapour pressure) may be taken as depending on the temperature, or the temperature (boiling point, melting point) as depending on the pressure.

Returning now to the general case expressed in equation (230), we find that the solution of foreign molecules, n_1, n_2, n_3, . . . affects the functional relation between T and p, which holds for the pure solvent. The deviation, in fact, depends only on the total number of dissolved molecules, and not on their nature. We can express this by taking either p or T as the independent variable. In the first case, we say that *at a definite pressure (p), the boiling point or the freezing point* (T) *of the solution differs from that of the pure solvent* (T$_0$); in the second case, *at a definite temperature* (T) *the vapour pressure, or the pressure of solidification (p) of the solution differs from that of the pure solvent* (p_0). We shall calculate the deviations in both cases.

§ 269. If the boiling point or the freezing point of the pure solvent at pressure p is T_0 then by (221a)

$$(\log K)_{T = T_0} = 0$$

and by subtracting from (230), we have

$$\log K - (\log K)_{T = T_0} = \frac{n_1 + n_2 + \cdots}{n_0}.$$

Since T differs only slightly from T_0, we can by Taylor's Theorem and equation (219) write :

$$\frac{\partial \log K}{\partial T} \cdot (T - T_0) = \frac{L}{RT_0^2}(T - T_0) = \frac{n_1 + n_2 + \cdots}{n_0}$$

Hence it follows that :

$$T - T_0 = \frac{n_1 + n_2 + \cdots}{n_0} \cdot \frac{RT_0^2}{L} \quad . \quad . \quad (231)$$

This law of the elevation of the boiling point, or of the depression of the freezing point, was first deduced by van't Hoff. When the process is one of freezing, L is negative. Since n_0 and L are multiplied together, nothing can be deduced about the number of molecules n_0 or about the molecular weight m_0 of the liquid solvent. If L is expressed in calories, we substitute 1·985 for R.

By the evaporation of 1 litre of water $n_0 L$ is approximately equal to 1000×539 cal., $T_0 = 373$, and therefore the elevation of the boiling point of a dilute aqueous solution is

$$T - T_0 = \frac{1 \cdot 985 \times 373^2}{1000 \times 539}(n_1 + n_2 + n_3 + \ldots),$$

$$= 0 \cdot 51°(n_1 + n_2 + n_3 + \ldots). \quad . \quad (232)$$

Further, by the freezing of 1 litre of water under atmospheric pressure, $n_0 L' = -1000 \times 80$ cal. approximately, $T_0' = 273$, and, therefore,

$$T_0' - T = \frac{1 \cdot 985 \times 273^2}{1000 \times 80}(n_1 + n_2 + n_3 + \ldots)$$

$$= 1 \cdot 85(n_1 + n_2 + n_3 + \ldots) \quad . \quad . \quad (233)$$

§ **270.** If the vapour pressure of the pure solvent at temperature T is p_0, then, by (221a),

$$(\log \mathrm{K})_{p = p_0} = 0$$

and, by subtracting from (230), we have

$$\log \mathrm{K} - (\log \mathrm{K})_{p = p_0} = \frac{n_1 + n_2 + \cdots}{n_0}.$$

Since p differs only slightly from p_0 we can, with the help of (220), write

$$\frac{\partial \log \mathrm{K}}{\partial p}(p - p_0) = -\frac{v}{\mathrm{RT}}(p - p_0) = \frac{n_1 + n_2 + \cdots}{n_0}.$$

Hence it follows, if v is put equal to the volume of the gas molecule which arises on the evaporation of a liquid molecule :

$$v = \frac{m_0}{m_0'}\,\frac{\mathrm{RT}}{p}$$

$$\frac{p_0 - p}{p} = \frac{m_0'}{m_0} \times \frac{n_1 + n_2 + \cdots}{n_0} \quad . \quad . \quad (234)$$

This law of the relative lowering of the vapour pressure was first deduced by van't Hoff. Since n_0 and m_0 occur only as a product, no conclusion can be drawn from the formula about the molecular weight of the liquid solution. This relation is frequently stated thus :—*The relative lowering of the vapour pressure of a solution is equal to the ratio of the number of the dissolved molecules* ($n_1 + n_2 + n_3 + \ldots$) *to the number of the molecules of the solvent* (n_0), *or, what is the same thing in dilute solutions, to the total number of the molecules of the solution.* This proposition holds only, as is evident, if $m_0 = m_0'$, *i.e.* if the molecules of the solvent possess the same molecular weight in the vapour as in the liquid. This, however, is not generally true, as, for example, in the case of water. It may be well therefore to emphasize this fact, that nothing concerning the molecular weight of the solvent can be inferred from the relative lowering of the vapour pressure, any more than from its boiling point, or freezing point. Measurements of this kind will not, under any circumstances, lead to anything but the total number $(n_1 + n_2 + \ldots)$ of the dissolved molecules.

§ 271. It is of interest to combine the special relations deduced in these last paragraphs with the corresponding equations which we deduced, in §§ 224, 226, and 228, for the same physical system, on general grounds, independent of any molecular theory.

We must then according to our present notation put for c of (162) :

$$c = \frac{M_2}{M_1} = \frac{n_1\, m_1 + n_2\, m_2 + \ldots}{n_0\, m_0} \quad . \quad . \quad (235)$$

Further, we must note that there m, the molecular weight of the gaseous solvent, is here denoted by m_0', and that there L, the heat per unit mass, is here denoted by L/m_0.

On carrying out the substitution, we see that in fact equations (181) and (183) are identical with (234) and (231), if we substitute for the characteristic positive quantity ψ the value

$$\psi = \frac{R(n_1 + n_2 + \ldots)}{n_1\, m_1 + n_2\, m_2 + \ldots} \quad . \quad . \quad (236)$$

§ 272. The value of the osmotic pressure p follows naturally from the general equation (190), if we substitute for the characteristic quantity ψ the value deduced in equation (236), and for c the expression (235)

$$P = \frac{RT}{n_0\, m_0\, v}(n_1 + n_2 + n_3 + \ldots) \quad . \quad . \quad (237)$$

Here v denotes, as formerly, the specific volume of the solution. The product $n_0\, m_0\, v$ is nearly equal to the total volume of the solution, V, and therefore

$$P = \frac{RT}{V}(n_1 + n_2 + n_3 + \ldots). \quad . \quad (238)$$

This is the well known relation of van't Hoff, which is

identical with the characteristic equation (16) of a mixture of ideal gases whose numbers of molecules are $n_1, n_2, n_3 \ldots$

Each of the theorems deduced in the preceding paragraphs contains a method of determining the total number of the dissolved molecules in a dilute solution. Should the number calculated from such a measurement disagree with the number calculated from the percentage composition of the solution on the assumption of normal molecules, some chemical change of the dissolved molecules must have taken place by dissociation, association, or the like. This inference is of great importance in the determination of the chemical nature of dilute solutions.

Raoult was the first to establish with exactness by experiment the relation between the depression of the freezing point and the number of the molecules of the dissolved substance; and van't Hoff gave a thermodynamical explanation and generalization of it by means of his theory of osmotic pressure. Application to electrolytes was rendered possible by Arrhenius' theory of electrolytic dissociation. Thermodynamics has led quite independently, by the method here described, to the necessity of postulating chemical changes of the dissolved substances in dilute solutions.

§ **273.** All the conclusions of the preceding paragraphs depend on the assumption of the equations (209), that the energy and the volume of a dilute solution are linear functions of the number of molecules. Many measurements of the influence of dilution on the freezing point and on the osmotic pressure of solutions of strong electrolytes have shown remarkable deviations from Ostwald's dilution law. It is easy to imagine that these assumptions are responsible for these deviations. In fact by introducing a generalization of these equations as suggested in § 252 on the lines adopted by J. C. Ghosh for the electrical conductivity of strong electrolytes we obtain a good agreement with the measurements on the lowering of the freezing point. This will be shown for the special case of a binary electrolyte, say KCl or NaCl.

According to J. C. Ghosh, all the dissolved molecules are completely dissociated into their ions. The system is therefore, according to (216) represented by

$$n_0 \; m_0, \; n_1 \; m_1, \; n_2 \; m_2 \mid n_0' \; m_0',$$

where n_0, n_0', m_0, and m_0' are the number of molecules and the molecular weights of the solvent (water) in the two phases in contact, $n_1 = n_2$, m_1 and m_2 are the number and weight of the ions in the dilute solution.

The concentrations of the molecules, neglecting vanishingly small quantities, are :

$$\left.\begin{aligned}
c_0 &= \frac{n_0}{n_0 + n_1 + n_2} = 1 - \frac{2n_1}{n_0} = 1 - 2c \\[2mm]
c_1 &= c_2 = \frac{n_1}{n_0 + n_1 + n_2} = \frac{n_1}{n_0} = c \\[2mm]
c_0' &= \frac{n_0'}{n_0'} = 1
\end{aligned}\right\} \quad (238a)$$

To determine the expression for the characteristic function Ψ of the solution we use the same method as in §§ 250 to 254, only with this difference, that we suppose that the coefficients of n_1 in the expression for the energy of the solution

$$U = n_0 \, u_0 + n_1(u_1 + u_2)$$

depend on the concentration $\dfrac{n_1}{n_0} = c$.

Since with vanishing c, or with n_0 increasing without limit, while n_1 remains constant, the correction term vanishes, the energy necessary to completely separate the ions, or the electric potential of the ions, will be approximately inversely proportional to the mean distance of two neighbouring oppositely charged ions, or, what evidently comes to the same thing, directly proportional to the cube root of the concentration c.

Instead of (209), we put

$$U = n_0 \, u_0 + n_1(\bar{u} - \beta c^{\frac{1}{3}}) \quad . \quad . \quad . \quad (238b)$$

where u still depends on temperature and pressure, while the positive quantity β is regarded as constant.

Accordingly, all further steps for determining Ψ are indicated. We may neglect all terms multiplied by p, since we may suppose the pressure as vanishingly small. For the entropy of the solution we get as in (211)

$$\Phi = n_0\phi_0 + n_1\bar{\phi} + C,$$

where

$$d\bar{\phi} = \frac{d\bar{u}}{T}$$

and

$$C = - R(n_0 \log c_0 + n_1 \log c_1 + n_2 \log c_2).$$

Finally, for the characteristic function of the solution, we get, instead of (215):

$$\Psi = \Phi - \frac{U}{T} = n_0\psi_0 - n_1\left(\bar{\psi} + \frac{\beta}{T}c^{\frac{1}{3}}\right) + C, \quad . \quad (238c)$$

where

$$\bar{\psi} = \bar{\phi} - \frac{\bar{u}}{T}$$

Thermodynamical equilibrium between the liquid and solid phases requires the condition :

$$\delta\Psi + \delta\Psi' = 0.$$

If further, as in (228), we put

$$\delta n_0 : \delta n_1 : \delta n_2 : \delta n_0' = - 1 : 0 : 0 : \frac{m_0}{m_0'},$$

and consider that :

$$\delta c = - \frac{n_1}{n_0^2}\delta n_0, \quad \delta C = - R \log c_0 . \delta n_0$$

$$- \psi_0 + \frac{1}{3} n_1 \frac{\beta}{T}c^{-\frac{2}{3}} \frac{n_1}{n_0^2} + R \log c_0 + \frac{m_0}{m_0'}\psi_0' = 0$$

or, by (238a),

$$\frac{1}{R}\left(\frac{m_0}{m_0'}\psi_0' - \psi_0\right) = \log K = 2c - \frac{1}{3} \frac{\beta}{RT}c^{\frac{1}{3}} \quad (238d)$$

This relation takes the place of equation (230) and gives for the change of the freezing point instead of the simple van't Hoff theorem (231) the expression

$$T - T_0 = 2c\left(1 - \frac{1}{6}\frac{\beta}{RT}c^{\frac{1}{2}}\right)\frac{RT^2}{L}, \quad . \quad . \quad (238e)$$

which is in agreement with the results of the measurements to date.*

It is easy to extend this theory to partially dissociated electrolytes. This generalisation shows complete analogy with the theory of the usual characteristic equation of a gas. Just as the deviations in the behaviour of a gas from that of an ideal gas may be ascribed to two quite different causes, to chemical actions (dissociation) or to physical actions (attractions between the unchanging molecules), so may the change, which the quotient of the lowering of the freezing point and the concentration of the dissolved molecules with progressive dilution shows, be caused either by chemical changes or by physical interaction between the dissolved molecules. According as one or the other or both together are active, we obtain for the influence of progressive dilution Ostwald's or Ghosh's or a more general law of dilution, which contains the other two as special cases.

§ 274. Each Phase contains both Constituents in Appreciable Quantity.—The most important case is the evaporation of a liquid solution, in which not only the solvent, but also the dissolved substance is volatile. The general equation of equilibrium (218), being applicable to mixtures of perfect gases whether the mixture may be supposed dilute or not, holds with corresponding approximation for a vapour of any composition. The liquid, on the other hand, must be assumed to be a dilute solution.

In general, all kinds of molecules will be present in both phases, and therefore the system is represented by

$$n_0\, m_0,\, n_1\, m_1,\, n_2\, m_2\, \ldots\, \mid\, n_0{}'\, m_0{}',\, n_1{}'\, m_1{}',\, n_2{}'\, m_2{}'\, \ldots$$

* Cf. Noyes and Falk, *J. Amer. Chem. Soc.*, **32**, 101, 1910; J. C. Ghosh, *Trans. Chem. Soc.*, **113**, 707, 790, 1918.

where, as hitherto, the suffix 0 refers to the solvent, the suffixes 1, 2, 3 . . . to the different kinds of the dissolved substances. By adding dashes to the molecular weights (m_0', m_1', m_2') we make it possible for the molecular weight of the vapour to be different from that of the liquid. The total number of molecules is, in the liquid,

$$n = n_0 + n_1 + n_2 + \ldots \text{ (nearly } = n_0),$$

in the vapour,

$$n' = n_0' + n_1' + n_2' + \ldots$$

The concentrations of the different kinds of molecules are, in the liquid,

$$c_0 = \frac{n_0}{n}, c_1 = \frac{n_1}{n}, c_2 = \frac{n_2}{n} \ldots$$

in the vapour,

$$c_0' = \frac{n_0'}{n'}, c_1' = \frac{n_1'}{n'}, c_2' = \frac{n_2'}{n'} \ldots$$

Since this system is capable of different kinds of changes, there are different conditions of equilibrium to be fulfilled, each of which refers to a definite kind of transformation. Let us first consider the reaction :

$$\nu_0 : \nu_1 : \nu_2 : \ldots : \nu_0' : \nu_1' : \nu_2' : \ldots$$
$$= \delta n_0 : \delta n_1 : \delta n_2 : \ldots : \delta n_0' : \delta n_1' : \delta n_2' : \ldots$$

which consists in the evaporation of a molecule of the first kind, and therefore

$$\nu_0 = 0, \nu_1 = -1, \nu_2 = 0, \ldots \nu_0' = 0, \nu_1' = \frac{m_1}{m_1'}, \nu_2' = 0, \ldots$$

The equation of equilibrium becomes

$$-\log c_1 + \frac{m_1}{m_1'} \log c_1' = \log K$$

or

$$\frac{c_1'^{\frac{m_1}{m_1'}}}{c_1} = K.$$

This equation enunciates Nernst's law of distribution. If the dissolved substance possesses the same molecular weight in both phases ($m_1' = m_1$), then there exists a fixed ratio of the concentrations, c_1 and c_1', in the liquid and vapour, which depends only on the pressure and temperature, but is independent of the presence of other kinds of molecules. If, however, the dissolved substance is somewhat polymerised, then the necessary relation of the last equation takes the place of the simple ratio.

If, on the other hand, a molecule of the solvent evaporates, we have,

$$\nu_0 = -1, \nu_1 = 0, \nu_2 = 0, \ldots \nu_0' = \frac{m_0}{m_0'}, \nu_1' = 0, \nu_2' = 0 \ldots;$$

and the equation of equilibrium becomes

$$- \log c_0 + \frac{m_0}{m_0'} \log c_0' = \log \mathrm{K},$$

where

$$- \log c_0 = \log \frac{n}{n_0} = \log\left(1 + \frac{n_1 + n_2 + \cdots}{n_0}\right) = \frac{n_1 + n_2 + \cdots}{n_0} \quad (239)$$

$$\therefore \qquad \frac{n_1 + n_2 + \cdots}{n_0} + \frac{m_0}{m_0'} \log c_0' = \log \mathrm{K}, \quad . \quad (240)$$

where $\frac{n_1}{n_0}, \frac{n_2}{n_0}, \ldots$, the concentrations of the molecules dissolved in the liquid, have small values. Two cases must be considered.

Either, the molecules m_0 in the vapour form only a small or at most a moderate portion of the number of the vapour molecules. Then the small numbers $\frac{n_1}{n_0}, \frac{n_2}{n_0}, \ldots$, may be neglected in comparison with the logarithm, and therefore

$$\frac{m_0}{m_0'} \log c_0' = \log \mathrm{K}.$$

An example of this is the evaporation of a dilute solution, when the solvent is not very volatile, *e.g.* alcohol in water.

Or, the molecules m_0 in the vapour far outnumber all the other molecules, as, *e.g.*, when alcohol is the solvent in the liquid phase, water the dissolved substance. Then c_0' is nearly equal to 1. We may make this simplification of the conditions of equilibrium but not without considerable error. Rather have we, as in (239),

$$\log c_0' = -\frac{n_1' + n_2' + \cdots}{n_0'}.$$

Equation (240) then becomes

$$\frac{n_1 + n_2 + \cdots}{n_0} - \frac{m_0}{m_0'} \times \frac{n_1' + n_2' + \cdots}{n_0'} = \log K.$$

If we deal with this equation exactly as we dealt with equation (230), we get finally, as in (231),

$$T - T_0 = \left(\frac{n_1 + n_2 + \cdots}{n_0 \, m_0} - \frac{n_1' + n_2' + \cdots}{n_0' \, m_0'}\right)\frac{RT^2 m_0}{L} \quad . \quad (240a)$$

Here L is the latent heat of evaporation of a gram molecule of the solvent, so $\dfrac{L}{m_0}$ is the latent heat of evaporation per unit mass.

We may again point out that the mass and not the number of molecules or molecular weight of the solvent enters into the formula. On the other hand, it is the molecular state of the dissolved substance that characterizes its influence on the evaporation. Besides, the formula contains a generalization of van't Hoff's law of the elevation of the boiling point in that here instead of the number of the dissolved molecules $(n_1 + n_2 + \cdots)$ the difference of the molecules dissolved in unit mass of the liquid and of the vapour occurs. The solution has a higher or lower boiling point than the solvent according as the liquid or the vapour contains the greater number of dissolved molecules. In the limit when both are equal, and the mixture boils at constant temperature, the change of the boiling point will be zero, as has already been shown from a general standpoint (§ 219). Naturally, corresponding laws hold for the change of the vapour pressure.

Exactly analogous theorems may, of course, be deduced for other states of aggregation. Thus, the more general statement of the law concerning the freezing point would be : *If both the solvent and the dissolved substance of a dilute solution solidify in such a way as to form another dilute solution, the depression of the freezing point is not proportional to the concentrations of the dissolved substances in the liquid, but to the difference of the concentrations of the dissolved substances in the liquid and solid phases, and changes sign with this difference.* The solidification of some alloys is an example. Eutectic mixtures, which do not change their composition on freezing, form the limiting case. The freezing point of the eutectic mixture does not depend on the concentration of the solution (cf. § 219b).

While these laws govern the distribution of the molecules in both phases, the equilibrium within each phase obeys the laws, which were deduced in § 262, etc. We again meet with the laws of dissociation, association, etc.

§ 275. Three Independent Constituents in one Phase.—Two dissolved substances in a dilute solution will not affect one another unless they form new kinds of molecules with one another, for there is no transformation possible, and therefore no special condition of equilibrium to fulfil. If two dilute aqueous solutions of totally different electrolytes which have no chemical action on one another be mixed, each solution will behave as if it had been diluted with a corresponding quantity of the pure solvent. The degree of the dissociation will rise to correspond to the greater dilution.

It is different when both electrolytes have an ion in common, as, for example, acetic acid and sodium acetate. In this case, before mixing there are two systems :

$$n_0 \, H_2O, \; n_1 \, CH_3.COOH, \; n_2 \, \overset{+}{H}, \; n_3 \, CH_3.\overset{-}{COO},$$

and $\quad n_0' \, H_2O, \; n_1' \, CH_3.COONa, \; n_2' \, \overset{+}{Na}, \; n_3' \, CH_3.\overset{-}{COO}.$

As in (222), for the first solution,

$$\frac{c_2{}^2}{c_1} = K, \text{ or } \frac{n_2{}^2}{n_1 n_0} = K, \quad . \quad . \quad . \quad (241)$$

for the second, $\quad \dfrac{c_2{}'^2}{c_1{}'} = K', \text{ or } \dfrac{n_2{}'^2}{n_1{}' n_0{}'} = K'. \quad . \quad . \quad (242)$

After mixing the two, we have the system

$$n_0 H_2O, \; n_1 CH_3.COOH, \; \bar{n}_2 CH_3.COONa, \; \overset{+}{\bar{n}_3 H}, \; \overset{+}{n_4 Na}, \bar{n}_5 \overset{-}{CH_3.COO},$$

where, necessarily,

$$\left. \begin{array}{l} \bar{n}_0 = n_0 + n_0{}' \text{ (number of } H_2O \text{ molecules)} \\ \bar{n}_2 + \bar{n}_4 = n_1{}' + n_2{}' \text{ (number of Na atoms)} \\ \bar{n}_1 + \bar{n}_3 = n_1 + n_2 \text{ (number of H atoms)} \\ \bar{n}_3 + \bar{n}_4 = \bar{n}_5 \text{ (number of free ions)} \end{array} \right\} \quad (243)$$

The total number of molecules in the system is

$$\bar{n} = \bar{n}_0 + \bar{n}_1 + \bar{n}_2 + \bar{n}_3 + \bar{n}_4 + \bar{n}_5 \text{ (nearly } = \bar{n}_0).$$

The concentrations are

$$\bar{c}_0 = \frac{\bar{n}_0}{\bar{n}}; \; \bar{c}_1 = \frac{\bar{n}_1}{\bar{n}}; \; \bar{c}_2 = \frac{\bar{n}_2}{\bar{n}}; \; \bar{c}_3 = \frac{\bar{n}_3}{\bar{n}}; \; \bar{c}_4 = \frac{\bar{n}_4}{\bar{n}}; \; \bar{c}_5 = \frac{\bar{n}_5}{\bar{n}}.$$

In the system there are two different reactions,

$$\nu_0 : \nu_1 : \nu_2 : \nu_3 : \nu_4 : \nu_5 = \delta n_0 : \delta \bar{n}_1 : \delta \bar{n}_2 : \delta \bar{n}_3 : \delta \bar{n}_4 : \delta \bar{n}_5,$$

possible; first, the dissociation of one molecule of acetic acid

$$\nu_0 = 0, \nu_1 = -1, \nu_2 = 0, \nu_3 = 1, \nu_4 = 0, \nu_5 = 1,$$

and therefore the condition of equilibrium is, by (218),

$$- \log \bar{c}_1 + \log \bar{c}_3 + \log \bar{c}_5 = \log K,$$

or $\quad \dfrac{\bar{c}_3 \, \bar{c}_5}{\bar{c}_1} = K, \text{ or } \dfrac{\bar{n}_3 . \bar{n}_5}{\bar{n}_1 . \bar{n}_0} = \dfrac{\bar{n}_3 . \bar{n}_5}{\bar{n}_1 (n_0 + n_0{}')} = K; \quad (244)$

second, the dissociation of a molecule of sodium acetate,

$$\nu_0 = 0, \nu_1 = 0, \nu_2 = -1, \nu_3 = 0, \nu_4 = 1, \nu_5 = 1,$$

whence, for equilibrium,

$$- \log \bar{c}_2 + \log \bar{c}_4 + \log \bar{c}_5 = \log \mathrm{K}', \text{ or } \frac{\bar{c}_4 \bar{c}_5}{\bar{c}_2} = \mathrm{K}',$$

or

$$\frac{\bar{n}_4 . \bar{n}_5}{\bar{n}_2 . \bar{n}_0} = \frac{\bar{n}_4 . \bar{n}_5}{\bar{n}_2(n_0 + n_0')} = \mathrm{K}'. \quad . \quad . \quad (245)$$

The quantities K and K' are the same as those in (241) and (242). They depend, besides on T and p, only on the nature of the reaction, and not on the concentrations, nor on other possible reactions. By the conditions of equilibrium (244) and (245), together with the four equations (243), the values of the six quantities $\bar{n}_0, \bar{n}_1, \ldots \bar{n}_5$ are uniquely determined, if the original solutions and also the number of molecules $n_0, n_1, \ldots$ and $n_0', n_1' \ldots$ be given.

§ 276. The condition that the two solutions should be *isohydric, i.e.* that their degree of dissociation should remain unchanged on mixing them, is evidently expressed by the two equations

$$\bar{n}_1 = n_1, \text{ and } \bar{n}_2 = n_1',$$

i.e. the number of undissociated molecules of both acetic acid and sodium acetate must be the same in the original solutions as in the mixture. It immediately follows, by (243), that

$$\bar{n}_3 = n_2, \bar{n}_4 = n_2', \bar{n}_5 = n_2 + n_2'.$$

These values, substituted in (244) and (245), and combined with (241) and (242), give

$$\frac{n_2(n_2 + n_2')}{n_1(n_0 + n_0')} = \mathrm{K} = \frac{n_2^2}{n_1 n_0},$$

$$\frac{n_2'(n_2 + n_2')}{n_1'(n_0 + n_0')} = \mathrm{K}' = \frac{n_2'^2}{n_1' n_0'},$$

whence the single condition of isohydric solutions is

$$\frac{n_2}{n_0} = \frac{n_2'}{n_0'}, \text{ or } c_2 = c_2'(= c_3 = c_3'),$$

or, *the two solutions are isohydric if the concentration of the common ion* $\overline{CH_3COO}$ *is the same in both.* This proposition was enunciated by Arrhenius, who verified it by numerous experiments. In all cases where this condition is not realized, chemical changes must take place on mixing the solutions, either dissociation or association. The direction and amount of these changes may be estimated by imagining the dissolved substances separate, and the entire solvent distributed over the two so as to form isohydric solutions. If, for instance, both solutions are originally normal (1 gram molecule in 1 litre of solution), they will not be isohydric, since sodium acetate in normal solution is more strongly dissociated, and has, therefore, a greater concentration of $\overline{CH_3}.COO$-ions, than acetic acid. In order to distribute the solvent so that the concentration of the common ion $\overline{CH_3}.COO$ may be the same in both solutions, some water must be withdrawn from the less dissociated electrolyte (acetic acid), and added to the more strongly dissociated (Na-acetate). For, though it is true that with decreasing dilution the dissociation of the acid becomes less, the concentration of free ions increases, as (262) shows, because the ions are now compressed into a smaller quantity of water. Conversely, the dissociation of the sodium acetate increases on the addition of water, but the concentration of the free ions decreases, because they are distributed over a larger quantity of water. In this way the concentration of the common ion $\overline{CH_3}.COO$ may be made the same in both solutions, and then their degree of dissociation will not be changed by mixing. This is also the state ultimately reached by the two normal solutions, when mixed. It follows, then, that when two equally diluted solutions of binary electrolytes are mixed, the dissociation of the more weakly dissociated recedes, while that of the more strongly dissociated increases still further.

§ 277. Three Independent Constituents in Two Phases.

—We shall first discuss the simple case, where the

second phase contains only one constituent in appreciable quantity. A solution of an almost insoluble salt in a liquid, to which a small quantity of a third substance has been added, forms an example of this case. Let us consider an aqueous solution of silver bromate and silver nitrate. This two-phase system is represented by

$$n_0 H_2O, n_1 AgBrO_3, n_2 AgNO_3, n_3 \overset{+}{Ag}, n_4 \overset{-}{BrO_3}, n_5 \overset{-}{NO_3} \mid n_0' AgBrO_3.$$

The concentrations are

$$c_0 = \frac{n_0}{n}; \; c_1 = \frac{n_1}{n}; \; c_2 = \frac{n_2}{n}; \ldots; \; c_0' = \frac{n_0'}{n_0'} = 1,$$

where $n = n_0 + n_1 + n_2 + n_3 + n_4 + n_5$ (nearly $= n_0$).

Of the possible reactions,

$$\nu_0 : \nu_1 : \nu_2 : \nu_3 : \nu_4 : \nu_5 : \nu_0' = \delta n_0 : \delta n_1 : \delta n_2 : \delta n_3 : \delta n_4 : \delta n_5 : \delta n_0',$$

we shall first consider the passage of one molecule of $AgBrO_3$ from the solution, viz.

$$\nu_0 = 0, \nu_1 = -1, \nu_2 = 0, \ldots \nu_0' = 1.$$

The condition of equilibrium is, therefore,

$$- \log c_1 + \log c_0' = \log K$$

or
$$c_1 = \frac{1}{K} \quad . \quad . \quad . \quad . \quad (246)$$

The concentration of the undissociated molecules of silver bromate in the saturated solution depends entirely on the temperature and the pressure, and is not influenced by the nitrate.

We may now consider the dissociation of a molecule of $AgBrO_3$ into its two ions.

$$\nu_0 = 0, \nu_1 = -1, \nu_2 = 0, \nu_3 = 1, \nu_4 = 1, \nu_5 = 0, \nu_0' = 0,$$

and, therefore,

$$- \log c_1 + \log c_3 + \log c_4 = \log K',$$

$$\frac{c_3 c_4}{c_1} = K',$$

or, by (246),
$$c_3 c_4 = \frac{K'}{K}, \quad . \quad . \quad . \quad . \quad (247)$$

i.e. the product of the concentrations of the $\overset{+}{Ag}$ and $\overset{-}{BrO_3}$ ions depends only on temperature and pressure. The concentration of the $\overset{+}{Ag}$-ions is inversely proportional to the concentration of the $\overset{-}{BrO_3}$-ions. Since the addition of silver nitrate increases the number of the $\overset{+}{Ag}$-ions, it diminishes the number of the $\overset{-}{BrO_3}$-ions, and thereby the solubility of the bromate, which is evidently measured by the sum $c_1 + c_4$.

We shall, finally, consider the dissociation of a molecule of $AgNO_3$ into its ions.

$$\nu_0 = 0, \nu_1 = 0, \nu_2 = -1, \nu_3 = 1, \nu_4 = 0, \nu_5 = 1, \nu_0' = 0,$$

whence, by (218),

$$\frac{c_3 c_5}{c_2} = K'' \quad . \quad . \quad . \quad . \quad . \quad (248)$$

To equations (246), (247), and (248), must be added, as a fourth, the condition

$$c_3 = c_4 + c_5,$$

and, as a fifth, the value of $c_2 + c_5$, given by the quantity of the nitrate added, so that the five unknown quantities, c_1, c_2, c_3, c_4, c_5, are uniquely determined.

The theory of such influences on solubility was first established by Nernst, and has been experimentally verified by him, and more recently by Noyes.

§ **278.** The more general case, where each of the two phases contains all three constituents, is realized in the distribution of a salt between two solvents, which are themselves soluble to a small extent in one another (*e.g.* water and ether). The equilibrium is completely determined by a combination of the conditions holding for the transition of molecules from one phase to another with those holding for the chemical reactions of the molecules within one and the same phase. The former set of conditions may be summed up in Nernst's law of distribution (§ 274). It assigns to each kind of molecule in the two phases a constant ratio of distribution, which is independent of the presence of other dissolved molecules. The second set is the conditions of the coexistence of three independent constituents in one phase (§ 275), to which must be added Arrhenius' theory of isohydric solutions.

§ 279. The same method applies to four or more independent constituents combined into one or several phases. The notation of the system is given in each case by (216), and any possible reaction of the system may be reduced to the form (217), which corresponds to the condition of equilibrium (218). All the conditions of equilibrium, together with the given conditions of the system, give the number of equations which the phase rule prescribes for the determination of the state of equilibrium.

When chemical interchanges between the different substances in solution are possible, as, *e.g.*, in a solution of dissociating salts and acids with common ions, the term *degree of dissociation* has no meaning, for the ions may be combined arbitrarily into dissociated molecules. For instance, in the solution

$$n_0 \; H_2O, \; n_1 \; NaCl, \; n_2 \; KCl, \; n_3 \; NaNO_3, \; n_4 \; KNO_3, \; n_5 \; \overset{+}{Na}, \; n_6 \; \overset{+}{K},$$
$$n_7 \; \overset{-}{Cl}, \; n_8 \; \overset{-}{NO_3}$$

we cannot tell which of the $\overset{+}{Na}$-ions should be regarded as belonging to NaCl, and which to $NaNO_3$. In such cases the only course is to characterize the state by the concentrations of the dissolved molecules.

The above system consists of water and four salts, but, besides the solvent, only three are independent constituents, for the quantities of the Na, the K, and the Cl determine that of the NO_3. Accordingly, by § 204 ($\alpha = 4, \beta = 1$) all the concentrations are completely determined at given temperature and pressure by three of them. This is independent of other kinds of molecules, and other reactions, which, as is likely, may have to be considered in establishing the conditions of equilibrium.

§ **280**. If in a system of any number of independent constituents in any number of phases, the condition of equilibrium (218) is not satisfied, *i.e.* if for any virtual isothermal isobaric change.

$$\sum \nu_0 \log c_0 + \nu_1 \log c_1 + \nu_2 \log c_2 + \ldots \gtrless \log K,$$

then the direction of the change which will actually take place in nature is given by the condition $d\Psi > 0$ (§ 147). If we now denote by $\nu_0, \nu_1, \nu_2 \ldots$, simple whole numbers, which are not only proportional to, but also of the same sign as the actual changes which take place, then we have, by (215),

$$\sum \nu_0 \log c_0 + \nu_1 \log c_1 + \nu_2 \log c_2 + \ldots < \log K, \quad (249)$$

for the direction of any actual isothermal isobaric change, whether it be a chemical change inside any single phase, or the passage of molecules between the different phases. The constant K is defined by (218).

To find the connection between the difference of the expressions on the right and left and the time of the reaction is immediately suggested. In fact, a general law for the velocity of an irreversible isothermal isobaric process may be deduced, by putting the velocity of the reaction proportional to that difference. We shall not, however, enter further into these considerations since the proportional factor cannot be determined without bringing in special atomic considerations.

CHAPTER VI.

ABSOLUTE VALUE OF THE ENTROPY.

NERNST'S THEOREM.

§ **281.** WE have repeatedly remarked that the whole thermo-dynamical behaviour of a substance is determined by a single characteristic function, a knowledge of which is sufficient once and for all to determine uniquely all the conditions of the physical and chemical states of equilibrium which the substance may assume. The form of the characteristic function depends, however, on the choice of the independent variables. If the energy U and the volume V are the independent variables, then the entropy is the characteristic function. If the temperature T and the volume V are the independent variables, then the characteristic function is the free energy, F. If the temperature T and the pressure p are the independent variables, then it is the function, Ψ (cf. § 152a).

The expression for the characteristic function may in each case be determined by the integration of certain quantities whose values can be determined by suitable measurement. In the expression for the entropy there remains, however, an additive constant a quite arbitrary, in that of the free energy a function of the form $a\mathrm{T} + b$ (§ 146), in that of the function Ψ one of the form $a + \dfrac{b}{\mathrm{T}}$ (cf. §§ 152B, 239). These additive terms can have no physical significance since only the differences of the energy, of the entropy, of the free energy etc. of the substance in different states play a part in natural processes. The additive terms compensate one another.

If the additive term is once fixed for any definite state

of a substance, it, of course, maintains that value for all other states of the substance. This is true not only for the same state of aggregation, and the same chemical modification, but also for other states of aggregation and other modifications, which the substance may assume.

Suppose that, after the value of the characteristic function has been determined for a definite arbitrarily selected modification of the substance, it is required to find it for some other modification, then we must search for a reversible physical transformation from one to the other which is measurable. This is necessary in order that the integration may be carried out along the path of the transformation. Such transformations are in many cases not realizable, since they often pass through regions of unstable states. As a rule, then, it is not possible to draw conclusions about the thermodynamical properties of one modification from that of another.

§ **282.** This gap in the theory has been filled by a theorem, which was developed by W. Nernst in 1906,* and which, since then, has been amply verified by experience. The gist of the theorem is contained in the statement that, as the temperature diminishes indefinitely the entropy of a chemical homogeneous (§ 67) body of finite density approaches indefinitely near to a definite value, which is independent of the pressure, the state of aggregation and of the special chemical modification.†

The interesting question as to the meaning of this theorem according to the molecular kinetic conception of the entropy cannot be answered here, since here we are concerned only

---

* *Nach. d. Ges. d. Wissensch. zu Göttingen, Math. phys. Kl.* 1906, Heft I. *Sitz.-Ber. d. preuss. Akad. d. Wiss.,* V. 20, Dezbr., 1906.

† This conception of the theorem is more comprehensive than that given by Nernst himselt. According to Nernst, the difference of the entropies of such a body in two different modifications is equal to zero in the limit when $T = 0$. This still leaves open the possibility of $lt_{T=0}$ entropy $= -\infty$. This distinction and others will be of practical importance when we come to discuss the value of the specific heat at the zero of absolute temperature (§ 284).

See a more general discussion by P. Gruner, *Verh. d. Deutschen Physik. Gesellschaft,* 14, S. 655, S. 727, 1912.

with general thermodynamics.* We must, however, deal with the important conclusions about physico-chemical equilibria to which it leads.

It is easy to see that since the value of the entropy contains an arbitrary additive constant, that, without loss of generality, we may write $lt_{T=0}$ entropy $= 0$. The Nernst heat theorem then states that, *as the temperature diminishes indefinitely, the entropy of a chemical homogeneous body of finite density approaches indefinitely near to the value zero.* The additive constant a of all chemical homogeneous substances in all states has herewith been uniquely disposed of. We can now in this sense speak of an absolute value of the entropy.

On the other hand, according to § 281, there remains in the value of the energy U, and of the free energy F of a body still an additive constant b, and in the value of the function Ψ an additive term of the form b/T undetermined, and arbitrary.

§ **283**. For the later applications of Nernst's heat theorem we must select the independent variables. We shall again, as in our former presentation, select the temperature T and the pressure p, since these quantities in the first place are easy to measure, and in the second place possess the advantage that in a system consisting of several phases they have the same values in all phases. With T and p as independent variables, it is necessary to choose Ψ as the characteristic function. For determining Ψ we have the general equation (150) of § 210,

$$\Psi = \Phi - \frac{H}{T} \quad \cdot \quad \cdot \quad \cdot \quad \cdot \quad (250)$$

where, by (150b),

$$\Phi = \int^T \frac{C_p dT}{T} \quad \cdot \quad \cdot \quad \cdot \quad \cdot \quad (251)$$

and

$$H = U + pV \quad \cdot \quad \cdot \quad \cdot \quad \cdot \quad (252)$$

If Ψ is known as a function of T and p, then, by § 152a,

---

* On the relation of Nernst's heat theorem to the quantum theory, see M. Planck, *Vorlesungen über die Theorie der Wärmestrahlung,* 4 Aufl., § 185.

the values of the heat function H, of the volume V, and of the entropy Φ are uniquely determined as functions of T and p :

$$H = T^2 \frac{\partial \Psi}{\partial T} \quad \cdots \quad \cdots \quad (253)$$

$$V = - T \frac{\partial \Psi}{\partial p} \quad \cdots \quad \cdots \quad (254)$$

$$\Phi = \Psi + T \frac{\partial \Psi}{\partial T} \quad \cdots \quad \cdots \quad (255)$$

In order to introduce Nernst's heat theorem into these quite general equations, let us consider a special system consisting of a chemical homogeneous solid or liquid body. The theorem is not directly applicable to a perfect gas, since perfect gases at the absolute zero of temperature do not possess a finite density. According to the theorem (§ 282), $\Phi = 0$, when $T = 0$.

It follows from (251) that

$$\int^0 \frac{C_p dT}{T} = 0.$$

Since both C_p and T are evidently positive, this equation is satisfied if, and always if, the lower limit of the integration is zero. Since the lower limit is independent of the temperature T, then

$$\Phi = \int_0^T \frac{C_p}{T} dT \quad \cdots \quad \cdots \quad (256)$$

holds for any temperature.

This equation is the mathematical expression for Nernst's heat theorem in its widest sense (cf. § 284). It follows from (250) that,

$$\Psi = \int_0^T \frac{C_p}{T} dT - \frac{H}{T} \quad \cdots \quad \cdots \quad (257)$$

or, on substituting the value of H from (150a),

$$\Psi = \int_0^T \frac{C_p}{T} dT - \frac{1}{T} \int^T C_p dT \quad \cdots \quad \cdots \quad (258)$$

The characteristic function Ψ for a chemically homogeneous solid or liquid body is hereby determined from measurements of the heat capacity C_p except for an additive term of the form b/T, which depends on the lower limit of the second integral. Although b is independent of the temperature, it may depend on the pressure p and on the chemical composition of the body. An additive constant remains arbitrary (cf. § 282).

§ **284.** Important conclusions with regard to the thermo-dynamical behaviour of solid and liquid substances at low temperatures may be drawn from the last equations. First of all from equation (256) the remarkable conclusion may be drawn, that the heat capacity of a chemically homogeneous solid or liquid body approaches indefinitely near to the value zero as the temperature is indefinitely decreased. Otherwise the entropy at finite temperatures could not possess a finite value.* Although this result is so striking at the first glance, yet measurement up to now have completely verified it.

Nernst's heat theorem also leads to an interesting conclusion with regard to the volume of a body. From equations (84b) and (256), we get:

$$\frac{\partial V}{\partial T} = -\frac{\partial \Phi}{\partial p} = -\int_0^T \frac{1}{T} \frac{\partial C_p}{\partial p} dT.$$

If we use equation (85),

then, $$\frac{\partial V}{\partial T} = \int_0^T \frac{\partial^2 V}{\partial T^2} dT = \frac{\partial V}{\partial T} - \left(\frac{\partial V}{\partial T}\right)_0$$

and therefore $$\left(\frac{\partial V}{\partial T}\right)_0 = 0,$$

---

* According to Nernst's conception the following equation,

$$\Phi' - \Phi = \int_0^T \frac{C_p' - C_p}{T} dT$$

takes the place of (256), where $\Phi' - \phi$ denotes the difference of the entropies of a solid or liquid body in two different modifications. From this we can only conclude that $C_p' = C_p$, when $T = 0$, but not that $C_p' = C_p = 0$. Should C_p and C_p' remain finite, then the entropy would not be equal to zero, when $T = 0$, but equal to $-\infty$. It is therefore of importance to note that should the conclusion $C_p = 0$ not be substantiated by experience, then Nernst's conception of the theorem could still be maintained as correct.

i.e., the coefficient of expansion of a chemically homogeneous solid or liquid body is in the limit, when T = 0, equal to zero. This conclusion is also in good agreement with experience, since the coefficient of expansion of most bodies show a marked decrease with decreasing temperature.

§ **285.** P. Debye * has deduced from the quantum hypothesis the following formula for the free energy of a solid body at low temperatures

$$F = U_0 - AT^4, \quad \ldots \quad (258a)$$

where U_0 is the energy at absolute zero, and the factor A depends only on the volume. It follows from (79a) that

$$p = -\frac{\partial F}{\partial V} = -\frac{\partial U_0}{\partial V} + T^4 \frac{\partial A}{\partial V}, \quad \ldots \quad (258b)$$

and
$$C_v = -T\frac{\partial^2 F}{\partial T^2} = 12AT^3, \quad \ldots \quad (258c)$$

in excellent agreement with the measurements.

The compressibility k has the value $-\dfrac{1}{V_0\left(\dfrac{\partial p}{\partial V}\right)_0}$ at absolute zero. Therefore the thermal coefficient of expansion at low temperatures is

$$\frac{1}{V_0}\left(\frac{\partial V}{\partial T}\right)_p = -\frac{1}{V_0}\frac{\left(\dfrac{\partial p}{\partial T}\right)_V}{\left(\dfrac{\partial p}{\partial V}\right)_T} = 4T^3\frac{\partial A}{\partial V} \times k_0 = \frac{C_v}{3A}\frac{\partial A}{\partial V}k_0.$$

The ratio of the coefficient of expansion to the specific heat is independent of temperature. This law was empirically deduced by E. Grüneisen.

§ **286.** Let us now consider two phases of the same chemically homogeneous solid or liquid substance in contact (*e.g.* at the melting point, or at the temperature of transformation

---

* P. Debye, "Wolfskehl-Vorträge über die Kinetische Theorie der Materie und der Elektrizität," Göttingen, 1914, p. 19 (Leipzig, Teubner).

of allotropic modifications). If n and n' denote the number of molecules of both bodies, then

$$\Psi = n\psi \qquad \Psi' = n'\psi'$$

where ψ and ψ' are the values per molecule m and m_1. They depend only on T, p, and the nature of the phase. We shall denote the entropy Φ, the energy U, the heat function H, and the volume V per molecule m by the corresponding small letters ϕ, u, h, v, without or with dashes, according as they belong to the first or second phase. On the other hand, to avoid confusion with the specific heat c_p, we shall denote the molecular heat by C_p.

The condition of equilibrium is, according to § 261 equation (221a),

$$\frac{\psi'}{m'} - \frac{\psi}{m} = 0$$

or, by (257)

$$\int_0^T \frac{1}{T}\left(\frac{C_p'}{m'} - \frac{C_p}{m}\right)dT - \frac{1}{T}\left(\frac{h'}{m'} - \frac{h}{m}\right) = 0$$

here

$$\frac{C_p'}{m'} - \frac{C_p}{m} = \Delta c_p \quad . \quad . \quad . \quad (259)$$

the difference of the specific heats, and

$$\frac{h'}{m'} - \frac{h}{m} = L,$$

the heat absorbed when unit mass passes from the phase without a dash to one with a dash by an isothermal isobaric process. We may write this shortly

$$\int_0^T \frac{\Delta c_p}{T}dT - \frac{L}{T} = 0 \quad . \quad . \quad . \quad (261)$$

or also, since by (150a) and (260)

$$\Delta c_p = \frac{\partial L}{\partial T} \quad . \quad . \quad . \quad . \quad (261a)$$

$$\int_0^T \frac{1}{T}\frac{\partial L}{\partial T}dT - \frac{L}{T} = 0 \quad . \quad . \quad . \quad . \quad (262)$$

This equation may be used to calculate the melting point, or the temperature of transformation of a substance, if the heat of the transformation, L, is known as a function of the temperature from heat measurements.

According to the measurements of Broensted, the heat of transformation of rhombic sulphur into monoclinic sulphur is given by

$$L = 1.57 + 1.15 \times 10^{-5}\,T^2 \text{ cal.}$$

approximately.

This formula, which, to be sure, must be inaccurate at low temperatures, since by (261a) it leads to a contradiction of Debye's law of specific heats (258c), gives at the temperature of transformation T

$$2.3 \times 10^{-5}\,T - \frac{1.57}{T} - 1.15 \times 10^{-5}\,T = 0$$

or
$$T = 369.5,$$

while measurement gave 368·4. For ice, at its melting point, we have

$$L = 80, \qquad T = 273.$$

Hence, from (261),

$$\int_0^{273} \frac{\Delta c_p}{T} dT = \frac{80}{273} = 0.293.$$

Now the difference of the specific heats of water and ice at 0° C. is

$$\Delta c_p = 1.00 - 0.51 = 0.49.$$

So this difference with decreasing temperature does not approach zero as a linear function of the temperature, but falls quicker and quicker with falling temperature, quite in the sense of the Debye Law.

§ **287.** While the entropy of solid and liquid chemically homogeneous bodies under any pressure converges to the limiting value zero at the absolute zero, the entropy of an ideal gas at low temperatures has only a meaning if the pressure p is less than the saturation pressure p_s of the gas at the temperature T. At higher pressures, the gaseous state is no longer stable, and also not perfect. Therefore we take $p < p_s$. For a perfect gas, by equation (195),

$$\Phi = n(C_p \log T - R \log p + k). \quad . \quad . \quad (263)$$

Hence, we have

$$\Phi > n(C_p \log T - R \log p_s + k). \quad . \quad . \quad (263a)$$

(See the end of § 288 about the value of p_s.) From the energy of an ideal gas, as given in equation (192),

$$U = n(C_v T + b) \quad . \quad . \quad . \quad . \quad (264)$$

and the volume

$$V = n\frac{RT}{p} \quad . \quad . \quad . \quad . \quad (265)$$

we have for the heat function,

$$H = U + pV = n(C_p T + b), \quad . \quad . \quad (266)$$

and therefore for the characteristic function of a chemically homogeneous ideal gas

$$\Psi = \Phi - \frac{H}{T} = n\Big(C_p \log T - R \log p + a - \frac{b}{T}\Big) = n\psi \ (267)$$

This formula, which we have already deduced, possesses here a new interest through the meaning of the constant a. For, according to § 282, this constant, which Nernst called the *chemical constant* of the particular gas, and which is of importance for all physical and chemical reactions of the gas, can be determined absolutely. In order to find it, we must transform the gas into the liquid or solid state along a reversible path, which lends itself to measurement. That can be most simply done by direct condensation. The search for the laws of a saturated vapour is the most direct way of determining the chemical constant a.

§ **288.** For the equilibrium of a chemically homogeneous liquid in contact with its chemically homogeneous vapour, we have

$$\frac{\psi'}{m'} = \frac{\psi}{m}$$

according to § 261, equation (221a).

If we substitute for ψ, which depends on the liquid molecule m, the value from equation (257) and for ψ', which depends on the gaseous molecule m', the value from equation (267), assuming that the vapour is a perfect gas, we have

$$C_p' \log T - R \log p + a - \frac{b}{T} = \frac{m'}{m}\left(\int_0^T \frac{C_p}{T}dT - \frac{h}{T}\right).$$

This equation shows how the pressure of the saturated vapour depends on the temperature. The constant b, together with the additive constant contained in h, may be expressed in terms of the latent heat of evaporation. The heat function is, by (266),

$$h' = C_p'T + b$$

per molecule of the vapour.

Also the latent heat of evaporation is per molecule of the vapour m',

$$L = h' - \frac{m'}{m}h = C_p' T + b - \frac{m'}{m}h \quad . \quad . \quad (268)$$

Substituting this in the last equation, we have,

$$C_p' \log T + C_p' - R \log p + a = \frac{L}{T} + \frac{m'}{m}\int_0^T \frac{C_p}{T}dT \quad (269)$$

At absolute zero the latent heat of evaporation L_0 is, by (268),

$$L_0 = b - \frac{m'}{m}h_0,$$

also

$$L - L_0 = C_p'T - \frac{m'}{m}(h - h_0),$$

and, by (150a),

$$L - L_0 = C_p'T - \frac{m'}{m}\int_0^T C_p dT.$$

This gives for the pressure p_s of a saturated vapour, on eliminating L from (269),

$$\log p_s = \frac{C_p'}{R} \log T - \frac{L_0}{RT} + \frac{a}{R} + \frac{m'}{mR}\left(\frac{1}{T}\int_0^T C_p dT\right.$$
$$\left. - \int_0^T \frac{C_p}{T}dT\right) \quad . \quad . \quad (270)$$

At very low temperatures we may neglect the integral and write for the pressure of a saturated vapour *

$$\log p_s = \frac{C_p'}{R} \log T - \frac{L_0}{RT} + \frac{a}{R} \quad . \quad . \quad (271)$$

Therefore, the chemical constant a of a chemically homogeneous vapour may be found from vapour-pressure measurements, particularly at low temperatures. The factor $\dfrac{C_p'}{R}$ in the first term is for monatomic gases nearly equal to $\dfrac{5}{2}$, since $\dfrac{C_p'}{C_v'} = \dfrac{5}{3}$ for monatomic gases, and $C_p' - C_v' = R$ always.

The methods of the quantum theory permit a direct calculation of the chemical constant from the molecular weight of the vapour without any particular measurement. This calculation cannot, however, be gone into here. If we compare (271) with (263a), we see that the entropy of a perfect gas assumes the value $+ \infty$ at the absolute zero of temperature.

§ **289.** The laws of saturated vapours may be derived in another way. We have seen by the application of the second law of thermodynamics (§ 172) that the pressure of a saturated vapour may be completely determined from the characteristic equation $p = f(T, v)$, of a homogeneous substance applicable to the liquid and the gaseous states. If v_1 and v_2 denote the specific volumes of the vapour and the liquid, p_1 and p_2 the corresponding pressures, then the pressure of the saturated vapour, as a function of the temperature, is given by the two equations

$$p_1 = p_2 \text{ and } \int_{v_2}^{v_1} p \, dv = p_1(v_1 - v_2).$$

The integration is to be carried out at constant temperature.

---

* Cf. W. Nernst, *Verh. d. Deutschen Phys. Ges.*, **11**, p. 313, 1909; **12**, p. 565, 1910. F. Pollitzer, "Die Berechnung chemischer Affinitäten nach dem Nernstschen Wärmetheorem," Stuttgart, F. Enke, 1912.

From this it follows at once that the chemical constant a can be calculated if the general characteristic equation of the substance is known, and that therefore the chemical constant must occur in the general characteristic equation. Already a number of forms of the characteristic equation have been proposed, besides that of van der Waals (12) and that of Clausius (12a), there is a series of others, which in certain limited regions fulfil their purpose with good approximation. None, as far as I see, is sufficiently comprehensive to be applicable to liquids at the lowest temperatures. Accordingly no equation gives a formula for the pressure of a saturated vapour which passes into equation (271) at low temperatures.

Van der Waals' equation, for instance, as can be shown by a simple calculation, gives the relation

$$\log p_s = -\log \frac{b^2}{a} - \frac{a}{bRT} \qquad . \quad . \quad (272)$$

for the pressure of the saturated vapour at low temperatures. The log T term is missing. This depends on the fact that according to van der Waals' equation the specific heat of the vapour converges towards the same value as that of the liquid in the limit when $T = 0$. This cannot be reconciled with Nernst's theorem. Another requirement of the theorem, which none of the proposed characteristic equations fulfils, is that the coefficient of expansion of a liquid substance vanishes when $T = 0$ (§ 285). Before we can hope to calculate the chemical constant from the characteristic equation, a form of that equation must be found which, when $T = 0$ and the pressure is finite and positive, is consistent with Nernst's theorem.

§ **290.** A mixture of perfect gases with the number of molecules $n_1, n_2, n_3 \ldots$ has, by (197), the entropy

$$\Phi = \sum n_1 (C_{p_1} \log T - R \log (c_1 p) + k_1), \quad . \quad (273)$$

the heat function, by (193) and (191),

$$H = U + pV = \sum n_1 (C_{p_1} T + b_1), \quad . \quad . \quad (274)$$

and the characteristic function, by (199) and (199a),

$$\Psi = \sum n_1 \Big(C_{p_1} \log T - R \log (c_1 p) + a_1 - \frac{b_1}{T} \Big), \quad . \quad (275)$$

where c_1, c_2, c_3, . . . denote the molecular concentrations

$$c_1 = \frac{n_1}{n_1 + n_2 + n_3 + \ldots}, \; c_2 = \frac{n_2}{n_1 + n_2 + n_3 + \ldots}, \ldots$$

The chemical constants a_1, a_2, a_3, . . . have particular values for each kind of molecule. They may be derived from the behaviour of each single chemical homogeneous component of the mixture, *e.g.*, from vapour pressure measurements. When this is done, we can, for each chemical reaction between the different kinds of molecules of the mixture,

$$\delta n_1 : \delta n_2 : \ldots = \nu_1 : \nu_2 : \ldots,$$

give immediately the characteristic constant of equation (201b)

$$\frac{\nu_1 a_1 + \nu_2 a_2 + \ldots}{R} = \log A.$$

Accordingly, the constants of the equilibrium condition (203a)

$$c_1{}^{\nu_1} c_2{}^{\nu_2} c_3{}^{\nu_3} \ldots = A e^{-\frac{B}{T}} T^C p^{-\nu} = K \quad . \quad (276)$$

are each derived from independent measurements. B is calculated, according to (205), from the heat of reaction, and C, according to (203), from the specific heat of each kind of molecule, while ν, according to (201a) denotes the increase of the number of the molecules caused by the reaction.

The constants a_1, a_2, . . . maintain their meaning, not only in reactions within the gaseous phase, but also in each reaction in which the gas mixture takes a part. The reaction may be combined with the precipitation of solid or liquid substances. This is true, since the expression (275) of the characteristic function governs quite generally the laws of thermodynamical equilibrium. We have only to substitute in the general equation (145), the values of Ψ obtained from equation (275) for a gaseous phase, and from equation

(257) or (258) for a chemically homogeneous solid or liquid phase.

§ **291.** Let us now consider the behaviour of solid or liquid solutions. Such a solution is, by (216), represented by

$$n_1 \, m_1, \; n_2 \, m_2, \; n_3 \, m_3, \; \ldots$$

The solution need not be dilute, and therefore the suffix 0 is dropped. This notation differs from our former notation.

In finding an expression for the characteristic function Ψ of the solution, we shall first of all investigate the entropy of the solution. The entropy of a solution is certainly not equal to zero, when $T = 0$, as it is for a chemically homogeneous body according to Nernst's heat theorem (§ 282). A glance at the expression (213) shows that the entropy of a dilute solution does not converge to zero as we approach the absolute zero. The expression contains additive terms, which are independent of the temperature. We may assume that the terms of Φ, which depend on the temperature, vanish when $T = 0$, like the entropy of a chemically homogeneous substance, and that, when $T = 0$,

$$\Phi = - R \sum n_1 \log c_1.$$

This will also hold for solutions of any dilution, since the expression $- R \sum n_1 \log c_1$ occurs as an additive term in values of the entropy, not only for small but for any values of the concentrations $c_1, c_2, \ldots$ independent of the state of aggregation, solid, liquid, or gaseous. This has been proved for perfect gases, equation (197), and we can always imagine solid and liquid bodies, by suitable changes of temperature and pressure, to pass continuously into the gaseous state, while the number of molecules n is kept constant.

Nernst's theorem applied to a chemically homogeneous body is expressed in the equation (256). We shall now extend the theorem to any liquid or solid solution with the molecular numbers $n_1, n_2, n_3 \ldots$ The entropy of such a solution is

$$\Phi = \int_0^T \frac{C_p}{T} dT - R \sum n_1 \log c_1, \quad . \quad . \quad . \quad (277)$$

where the integration with respect to T has to be carried out at constant pressure and constant n. The characteristic function Ψ of the solution is, therefore,

$$\Psi = \int_0^T \frac{C_p}{T} dT - R \sum n_1 \log c_1 - \frac{H}{T} \quad . \quad . \quad (278)$$

and the remaining thermodynamical properties of the solution are according to § 283 uniquely determined. When the substance is chemically homogeneous, the equations, of course, are the same as those already derived, since then $c_1 = 1, c_2 = 0, c_3 = 0 \ldots$.

In the special case of a dilute solution, of course, all the relations, which were deduced in Chapter V, hold good. To these must now be added the conclusion, which comes readily from equation (277), that the terms ϕ_1, ϕ_2, ϕ_3 . . ., which appear in the expression for the entropy of a dilute solution, vanish when $T = 0$.

§ 292. From equation (277) it follows directly, just as in §§ 284 and 285 from equation (256) in the case of a chemically homogeneous body, that *both the heat capacity and the coefficient of expansion of each solid and liquid body converges towards the value zero when the temperature is indefinitely decreased.*

§ 293. If a system consist of any number of solid and liquid phases, then the system is represented by

$$n_1 m_1, n_2 m_2, \ldots \mid n_1' m_1', n_2' m_2' \ldots \mid n_1'' m_1'', n_2'' m_2'' \ldots \mid \ldots$$

The general condition of equilibrium

$$\sum \delta \Psi = 0$$

holds good for a possible isothermal and isobaric transformation of the kind

$$\delta n_1 : \delta n_2 : \ldots : \delta n_1' : \delta n_2' : \ldots : \delta n_1'' : \delta n_2'' : \ldots$$
$$= \nu_1 : \nu_2 : \ldots : \nu_1' : \nu_2' : \ldots : \nu_1'' : \nu_2'' : \ldots$$

Also

$$\sum \frac{\partial \Psi}{\partial n_1} \nu_1 + \frac{\partial \Psi}{\partial n_2} \nu_2 + \ldots = 0, \quad . \quad . \quad (279)$$

where the summation extends over all the phases. If we substitute the value of Ψ from (278), we see that the equilibrium evidently depends on the values of $\dfrac{\partial C_p}{\partial n_1}, \dfrac{\partial C_p}{\partial n_2}, \ldots$ and $\dfrac{\partial H}{\partial n_1}, \dfrac{\partial H}{\partial n_2} \ldots$ *i.e.*, it depends on how the heat function H, and the heat capacity C_p, of each phase vary with the number of molecules. C_p is, of course, determined through H, since by (150*a*).

$$C_p = \frac{\partial H}{\partial T}.$$

Accordingly, the determination of the laws of chemical equilibrium is made to depend on measurements of heat capacity and heat of reaction. To be sure, the experimental results available are not sufficiently extensive to test completely this far-reaching conclusion of the theory.

To shorten the notation, we shall write

$$\frac{\partial H}{\partial n_1} = h_1, \frac{\partial H}{\partial n_2} = h_2, \quad . \quad . \quad . \quad (280)$$

and therefore by (150*a*)

$$\frac{\partial C_p}{\partial n_1} = \frac{\partial h_1}{\partial T}, \frac{\partial C_p}{\partial n_2} = \frac{\partial h_2}{\partial T} \quad . \quad . \quad (281)$$

The conditions of equilibrium of the whole system may then, with the help of (278), be written

$$\sum \nu_1 \left(\int_0^T \frac{\partial h_1}{\partial T} \frac{dT}{T} - \frac{h_1}{T} - R \log c_1 \right)$$
$$+ \nu_1 \left(\int_0^T \frac{\partial h_2}{\partial T} \frac{dT}{T} - \frac{h_2}{T} - R \log c_2 \right) + \ldots = 0.$$

Now

$$\sum \nu_1 h_1 + \nu_2 h_2 + \ldots = L$$

is the heat absorbed when the isothermal and isobaric transformation represented by ν takes place. Hence the conditions of equilibrium is more simply represented by

$$\int_0^T \frac{\partial L}{\partial T} \cdot \frac{dT}{T} - \frac{L}{T} - R \left(\sum \nu_1 \log c_1 + \nu_2 \log c_2 + \ldots \right) = 0$$

or

$$\sum \nu_1 \log c_1 + \nu_2 \log c_2 + \ldots$$
$$= \frac{1}{R} \left(\int_0^T \frac{\partial L}{\partial T} \cdot \frac{dT}{T} - \frac{L}{T} \right) = \log K \quad . \quad . \quad (282)$$

This form of the equation is in complete agreement with the condition of equilibrium (218) of a system of dilute solutions. There is, however, an evident difference. First of all, the quantity log K appears here to be completely determined from the heat of reaction L without an unknown additive constant.* In the second place, the quantity K, as well as L, depends, besides on T and p, also on the finite concentrations $c_1, c_2, \ldots$ This was not the case before. This arises from the fact that here the equilibrium concentrations cannot be expressed directly as functions of T and p, as is possible with dilute solutions. Since the quantity K is finite, $\dfrac{\partial L}{\partial T}$ must vanish when T = 0. From this it follows that *the heat of a reaction between solid and liquid bodies has no temperature coefficient at absolute zero.* This law forms an important preliminary condition which each empirical formula must fulfil if the formula is to be used for determining K.

Instead of equation (282), we may often with advantage use the exactly equivalent relation

$$\sum \nu_1 \log c_1 + \nu_2 \log c_2 + \ldots$$

$$= \frac{1}{R}\left(\int_0^T \frac{L - L_0'}{T^2}dT - \frac{L_0}{T}\right) = \log K, \quad . \quad (283)$$

where L_0 denotes the heat of reaction at T = 0.

Log K is evidently infinite when T = 0. This is in complete agreement with the conclusion, which we drew in § 259A about dilute solutions, that at absolute zero all reactions with a finite heat value proceed until the reactions are completely finished.

§ **294.** The relations (219) and (220)

$$\frac{\partial \log K}{\partial T} = \frac{L}{RT^2} \text{ and } \frac{\partial \log K}{\partial p} = -\frac{v}{RT} \quad . \quad (284)$$

hold here also. This is found directly by differentiating the

---

* The integration with respect to T is, of course, to be carried out at constant pressure and constant concentration.

equation (282) with respect to T and p, remembering that the change of volume v of the system is given by the general formula (79g)

$$\frac{\partial L}{\partial p} = -T^2\frac{\partial}{\partial T}\left(\frac{v}{T}\right) = v - T\frac{\partial v}{\partial T} \quad . \quad . \quad (285)$$

Equation (282) gives the absolute value of the equilibrium constant K, while (284) leaves the integration constant still undetermined. If, for example, L = 0, then it follows from (284) that K is independent of the temperature, but its value remains undetermined. On the other hand, from (282), log K = 0. Therefore a solution of two enantiomorphic forms of an optically active compound can be in stable equilibrium only if it forms a racemic, optically inactive mixture, a conclusion peculiar to the Nernst theorem. Without this theorem we can only conclude that the composition of the mixture does not change with temperature.

§ 295. Calculation of the Degree of Dissociation of an Electrolyte from the Heat of Dissociation.—

If we take a solution of an electrolyte *e.g.*, acetic acid in water, then the system, as in § 262, is represented by

$$n_1\ H_2O,\ n_2\ CH_3.COOH,\ n_3\overset{+}{H},\ n_4\ CH_3.\overset{-}{COO}.$$

The total number of molecules is

$$n = n_1 + n_2 + n_3 + n_4.$$

The concentrations are

$$c_1 = \frac{n_1}{n},\ c_2 = \frac{n_2}{n},\ c_3 = \frac{n_3}{n},\ c_4 = \frac{n_4}{n}.$$

The dissociation of a molecule of acetic acid gives

$$\nu_1 = 0,\ \nu_2 = -1,\ \nu_3 = 1,\ \nu_4 = 1.$$

Therefore in equilibrium, by (283), since $c_3 = c_4$,

$$R\log\frac{c_3^2}{c_2} = R\log K = \int_0^T\frac{L - L_0}{T^2}dT - \frac{L_0}{T},$$

c_2 and c_3 can each be calculated from this, since the sum $c_2 + c_3$ is the total sum of the undissociated and dissociated acid molecules in the solution. Of course, the equilibrium formula can be used in general only if we know how the heat of dissociation depends on the temperature as well as on the concentration. With dilute solutions it is sufficient to know the temperature coefficient, since L may be taken as nearly independent of the concentration.

§ 296. Calculation of the Solubility from the Heat of Solution.—Let us consider the equilibrium of a salt solution in contact with a solid (ice, salt). The system is represented by

$$n_1\, m_1,\, n_2\, m_2 \mid n_1'\, m_1'.$$

The suffix 1 refers to the substance present in both phases, the suffix 2 to the substance present in the solution. The concentrations in the two phases are:

$$c_1 = \frac{n_1}{n_1 + n_2},\, c_2 = \frac{n_2}{n_1 + n_2},\, c_1' = \frac{n_1'}{n_1'} = 1.$$

Here we cannot, of course, speak of a definite solvent, since both substances in the solution may occur in any proportion.

When one molecule of the liquid is precipitated, the values are:

$$\nu_1 = -1,\, \nu_2 = 0,\, \nu_1' = \frac{m_1}{m_1'}.$$

Therefore, by (283), the condition of equilibrium is,

$$- R \log c_1 = \int_0^T \frac{L - L_0}{T^2} dT - \frac{L_0}{T}.$$

— L is the heat set free by the precipitation of a molecule of the liquid. This equation gives the value of the concentration c_1, and accordingly of the ratio $n_1 : n_2$, if L is known as a function of the temperature and the concentration. We may also draw a conclusion as to the molecular state of both substances in the solution.

§ **297.** If in the system considered there be, besides any number of solid and liquid phases, also a gaseous phase, and if, as formerly, we suppose this phase to be a perfect gas, the condition of equilibrium (283) may be easily, from equation (79), with the help of (275), generalized to

$$\left. \begin{aligned} &\sum \nu_1 \log c_1 + \nu_2 \log c_2 + \ldots \\ &= \frac{1}{R} \int_0^T \frac{L'}{T^2} dT - \frac{L_0}{RT} + C \log T - \nu \log p + \log A \\ &\qquad\qquad = \log K. \end{aligned} \right\} \quad (286)$$

where, for shortness,

$$L' = L - L_0 - RCT \quad . \quad . \quad . \quad (287)$$

The constants A, C and ν refer, as in (276), to the gas phase alone.

This last equation (286) *contains all the conditions of equilibrium established in the last three chapters as special cases* (*changes in the state of aggregation, solubility, dissociation, lowering of the freezing point, raising of the boiling point, etc.*). Let us therefore, in conclusion, once more summarize the meaning of the symbols used therein.

On the left, the summation $\sum$ extends over all the phases of the system. $c_1, c_2, c_3 \ldots$ denote the molecular concentrations of the individual kinds of molecules with the number of molecules $n_1, n_2, n_3, \ldots$ in each phase :

$$c_1 = \frac{n_1}{n_1 + n_2 + n_3 + \ldots}, \; c_2 = \frac{n_2}{n_1 + n_2 + n_3 + \ldots},$$

$\nu_1, \nu_2, \nu_3 \ldots$ are the simultaneous changes of the molecular numbers $n_1, n_2, n_3 \ldots$ in any isothermal and isobaric transformation of the system, should it take place in any single phase, or in the passage of molecules from one phase to another phase.

On the right, R denotes the absolute gas constant, L' the expression (287), L the total heat of reaction of an isothermal and isobaric transformation, positive, if heat is absorbed, L_0 the value of L, when $T = 0$. The constants A, C, and ν refer to the gaseous phase alone, if there is one. According to (201a), ν is the change of the total number of gaseous molecules produced by the transformation. A is found from the individual kinds of gas molecules by (201b), and C from the specific heat by (203). The integration with respect to T is to be carried out at constant pressure and constant concentration.

In order that the temperature integral may have a finite value in spite of the lower limit 0, L′ as well as $\dfrac{\partial L'}{\partial T}$ must vanish, when $T = 0$. The first can be seen directly from (287), the latter can be understood, if we remember that $\dfrac{\partial L}{\partial T}$, by § 105, is in general equal to the difference of the heat capacities of the system after and before the transformation. This difference vanishes at $T = 0$, when the phases are solid or liquid, but, by (203), is equal to RC, when the phase is gaseous.

Often it is of advantage to write the temperature integral in the form

$$\int_0^T \frac{L'}{T^2}dT = -\frac{L'}{T} + \int_0^T \frac{\partial L'}{\partial T}\frac{dT}{T}$$
$$= -\frac{L - L_0 - RCT}{T} + \int_0^T\left(\frac{\partial L}{\partial T} - RC\right)\frac{dT}{T};$$

since $\dfrac{\partial L}{\partial T}$ may be determined from measurements of heat capacities.

When the system is condensed, ν, C, and log A are equal to zero. When it is a single gaseous phase, then, by (205), $L' = 0$. When it is a chemically homogeneous body, the terms with the concentrations c_1, c_2, . . . vanish.

In each special case the simple equations, which were formerly derived, result.

INDEX.

The numbers refer to pages.